Sixth

Edition

QUANTITATIVE

ANALYSIS

LABORATORY MANUAL

R. A. DAY, JR.
Emory University, Emeritus

A. L. UNDERWOOD
Emory University

PRENTICE-HALL, ENGLEWOOD CLIFFS, NEW JERSEY 07632

Editorial/production supervision: *Robert C. Walters*
Cover designer: *William Frost*
Prepress buyer: *Paula Massenaro*
Manufacturing buyer: *Lori Bulwin*
Supplements acquisitions editor: *Alison Muñoz*
Acquisitions editor: *Daniel Joraanstad*

 © 1991 by Prentice-Hall, Inc.
A Simon & Schuster Company
Englewood Cliffs, New Jersey 07632

Printed in the United States of America

10 9 8 7 6 5 4 3 2 1

0-13-747403-2

Prentice-Hall International (UK) Limited, *London*
Prentice-Hall of Australia Pty. Limited, *Sydney*
Prentice-Hall Canada Inc., *Toronto*
Prentice-Hall Hispanoamericana, S.A., *Mexico*
Prentice-Hall of India Private Limited, *New Delhi*
Prentice-Hall of Japan, Inc., *Tokyo*
Simon & Schuster Asia Pte. Ltd., *Singapore*
Editora Prentice-Hall do Brasil, Ltda., *Rio de Janeiro*

Contents

LABORATORY PROCEDURES

Preface

This laboratory manual has been written to accompany the sixth edition of the authors' *Quantitative Analysis* textbook. This text now contains a section on laboratory procedures suitable for a one-semester course. The manual contains considerably more experiments than does the text, and the number of topics is greater. It contains sufficient material for courses that offer two semesters of laboratory. Previous editions of the manual have also been used in courses that teach analysis in a laboratory format only, as well as in courses for students who do not expect to become professional chemists.

The basic format of the manual remains the same. After introductory chapters on general laboratory directions and the analytical balance, directions are given for a large number of experiments which illustrate a wide variety of analytical principles. Many experiments from the classical areas of titrimetric and gravimetric methods of analysis are included, as well as a number that may be termed "simple instrumental" methods. The latter experiments use relatively simple instruments, such as the Spectronic 20 and the pH meter. A few experiments are included which involve more expensive instruments, such as UV, IR, and atomic absorption spectrophotometers, a flame photometer, and gas and liquid chromatographs.

We have retained in several chapters some "descriptive" chemistry that we hope will make the experiments more relevant and interesting to the student. The chapter on the analytical balance has been extensively revised to give coverage to the now widely used electronic balances.

We wish to express our appreciation to many users of the fifth edition who have made numerous helpful suggestions. Our particular thanks go to Professor Hubert L. Youmans, who has given us permission to include several of his experiments. We are also grateful to Professor Harvey Diehl for permission to use his experiment illustrating the Craig apparatus, and to Professors Ronald C. Johnson and Stanley N. Deming for many helpful comments.

R. A. Day, Jr.
A. L. Underwood

1

General Laboratory Directions

INTRODUCTION

In determining the constitution of an unknown sample, a sound method must first be selected. It then remains a matter of technique alone whether the ultimate measurement is performed upon *all* or part of the desired constituent. The heart of *quantitative technique* is simply to carry a sample through a number of manipulations without accidental losses and without introducing foreign material. Since every conceivable fortuity cannot be anticipated in writing a text, the student must develop independent judgment in connection with his laboratory work. Common sense plus awareness of the danger spots are the main requirements of the beginning student in this regard.

Neatness and Cleanliness

Good analysts are usually scrupulously neat. The student with an orderly laboratory bench is not likely to mix up samples, add the wrong reagents, or spill solutions and break glassware. Neatness in the laboratory must extend, of course, from the student's own bench to the shelves where materials are available for the whole class. Much time can be wasted in searching for a small item in a jumble of glassware or in finding a certain reagent bottle that has been misplaced on the side-shelf. Neatness also includes stewardship over the more permanent laboratory fixtures, such as ovens, hot plates, hoods, sinks, and the benches themselves. Corrosive materials that are spilled must be cleaned up immediately from equipment, benches, or floors. It is important that plumbing be conserved by washing acids and bases down the drains with copious volumes of water. Because of environmental concerns and a heightened appreciation for toxicology, chemists are restricted in using public sewers for waste disposal; the student who believes that pure water is a desirable community goal will not subvert local laboratory regulations for disposal of chemicals.

No analysis should ever be performed using anything but clean glassware. Glassware that looks clean may or may not be clean as an analyst understands the term. Surfaces on which no visible dirt appears are often still contaminated by a thin, invisible film of greasy material. When water is delivered from a vessel so contaminated, it does not drain uniformly from the glass surface, but leaves behind isolated drops that are troublesome or sometimes impossible to recover. Glassware into which a brush can be inserted, such as beakers or Erlenmeyer flasks, are best cleaned with detergent. Pipets, burets, and volumetric flasks, on the other hand, are ordinarily cleaned with *cleaning solution*,[1] whose strong oxidizing properties ensure clean glass surfaces in most cases. After cleaning, apparatus should be rinsed several times with tap water, then with small portions of distilled water, and finally allowed to drain.

Planning and Efficiency

Beginning students will find that time spent in planning their work will save them many wasted hours during the laboratory periods. Before arriving at the laboratory, they should be familiar with the experiment to be performed. Some of the operations of analytical chemistry are very time consuming. Fortunately, however, some of these operations require very little attention once set in motion. For example, it is sometimes necessary to dry primary standards or unknown samples in an oven before they are used. It is foolish for students to come to the laboratory, place material in the oven, and then sit around for an hour or more while it dries. Depending upon the laboratory regulations in the particular school, it may be possible for them to place a sample in the oven before the regular laboratory period, or, perhaps, to work on another experiment while a sample is drying during the laboratory period. Filtration and ignition of precipitates are examples of lengthy operations that do not need constant attention. Students should plan their work, considering more than one experiment at a time (with the instructor's help at first, if necessary) in such a way that they need not be idle for long periods.

---

[1] Many different formulas are available for the preparation of cleaning solution, and the choice seems largely personal. A satisfactory solution can be made as follows: In a 600-ml beaker place 20 to 25 g of technical-grade sodium dichromate and 15 ml of water. Then add slowly and carefully about 450 ml of technical-grade sulfuric acid with occasional stirring. Cool the solution and store it in a glass-stoppered bottle. The appearance of the green chromium (III) ion indicates that the solution is exhausted. Cleaning solution is very corrosive; take special precautions not to get it on your skin or clothing. If any is spilled on the skin, rinse the affected parts quickly and thoroughly with tap water. Consult your instructor for directions before using this solution. Cleaning solution is not discarded after use. It is returned to the storage bottle until the color turns green.

Reagents

The purity of the reagents used in analytical chemistry is a matter of utmost importance. Fortunately, reagents sufficiently pure for most analytical work are now available commercially, and even primary standards can be purchased in some cases. In work where a certain impurity might be very deleterious, however, it is best to test each batch of reagents rather than to rely on the manufacturer. Except in special cases as noted in the laboratory directions, the student in the beginning course may assume that the reagents provided are of adequate purity.

Reagents of high purity cost more, of course, than reagents that need not be so carefully manufactured. For some uses, less expensive grades of chemicals may be employed. For example, the most costly grades of sulfuric acid and alkali dichromate would be extravagant for the preparation of "cleaning solution." Also, many examples can be found of cheaper grades of reagents which do not contain the particular impurities that imperil certain analyses. However, since a reagent is often used for a number of different analyses, it is generally most practical to stock the analytical laboratory with only the better grades of reagents.

Reagents are roughly classified as follows, although some of the designations have not been precisely defined:

1. *Technical (sometimes called practical or commercial)-grade* chemicals are used mainly in industrial processes on a large scale, and are seldom employed in the analytical laboratory except for such purposes as the preparation of cleaning solution.

2. *U.S.P.* regeants meet purity standards that can be found in the *United States Pharmacopoeia*. These standards were established primarily for the guidance of pharmacists and the medical profession, and in many cases, impurities are freely tolerated that are not incompatible with the use to which these persons will put the compounds. Thus U.S.P. reagents are not usually suitable for analytical chemistry.

3. *C.P.* reagents are often much more pure than U.S.P. reagents. On the other hand, the designation C.P. (standing for "chemically pure") has no definite meaning; standards of purity for reagents of this class have not been established. Thus C.P. reagents may often be used for analytical purposes, but there are many situations where they are not sufficiently pure. In many analyses it is necessary to test these reagents for certain impurities before they may be used.

4. *Reagent-grade* chemicals conform to the specifications established by the Committee on Analytical Reagents of the American Chemical Society.[2] The label on such a reagent bears a statement such as "meets A.C.S. specifications," and generally furnishes information regarding the actual percentages of various impurities or at least maximal limits of impurities.

[2] *Reagent Chemicals, American Chemical Society Specifications*, 5th ed., American Chemical Society, Washington, DC, 1974.

TABLE 1.1 LABEL OF A REAGENT-GRADE CHEMICAL

Potassium Biphthalate, Crystal, Primary Standard	
Meets A.C.S. specifications	
Assay	99.95–100.05%
Insoluble matter	0.005%
pH of a 0.05 M solution at 25°C	4.00
Chlorine compounds (as Cl)	0.003%
Sulfur compounds (as S)	0.002%
Heavy metals (as Pb)	0.0005%
Iron (Fe)	0.0005%
Sodium (Na)	0.0005%

Many reagent-grade chemicals are used as primary standards in the introductory laboratory. A label from a bottle of potassium acid phthalate (potassium biphthalate) is shown in Table 1.1. Note that the purity is 99.95 to 100.05%, well within the usual requirements for a primary standard.

Substances of even higher degrees of purity are available for research and work in such areas as solid-state physics. These ultrapure materials may be obtained from the National Institute of Standards and Technology and various commercial supply houses.

Certain acids and bases are provided in the laboratory as concentrated solutions. Table 1.2 gives the composition of most of the common reagents. Dilute solutions are prepared as needed by adding the concentrated solution to water. The degree of dilution is often indicated by the ratio of the volume of concentrated solution to that of water. For example, 1:4 nitric acid means that 1 volume of nitric acid is added to 4 volumes of water.

The chances of contamination increase enormously when a reagent bottle is placed in the laboratory for the use of a large number of people. Thus it is most important that students carefully adhere to certain rules governing the use of the reagent shelf. In addition to the following instructions, any further suggestions of the instructor should be heeded. (1) The reagent shelf should be clean and orderly.

TABLE 1.2 COMPOSITION OF CONCENTRATED COMMON ACIDS AND BASES

Reagent	Density, g/ml	Percent by Weight	Approximate Molarity
Acetic acid, $HC_2H_3O_2$	1.05	99.7	17.4
Ammonium hydroxide, NH_4OH	0.90	29 (as NH_3)	15.3
Hydrochloric acid, HCl	1.19	37	12.0
Nitric acid, HNO_3	1.42	70.5	15.9
Perchloric acid, $HClO_4$	1.68	71	11.9
Phosphoric acid, H_3PO_4	1.71	86	15.0
Sodium hydroxide, NaOH	1.54	51	19.6
Sulfuric acid, H_2SO_4	1.84	96.5	18.1

(2) Any spilled chemicals must be cleaned up immediately. (3) The stoppers of reagent bottles should not be placed on the shelf or laboratory bench. Stoppers may be placed on clean towels or watch glasses, although it is best to hold them between two fingers while reagents are being withdrawn. (4) The mouths of reagent bottles should be kept clean. (5) Pipets, droppers, or other instruments should never be inserted into reagent bottles. Rather, a slight excess of reagent should be poured into a clean beaker from which the pipetting is done and the excess discarded, not returned to the bottle. (6) Fingers, spatulas, or other implements should not be inserted into bottles of solid reagents.

APPARATUS

In addition to the usual equipment found in any chemistry laboratory, there are certain items that are of special interest to the analytical chemist. Some of the more important items are described in this section, and advice is given regarding their use.

Wash Bottle

Each student should have a wash bottle of reasonable capacity, capable of delivering a stream of distilled water from a tip connected to the main part of the bottle. A convenient type, shown in Figure 1.1(a), is made from polyethylene, and the body is squeezed to force water from the tip. The wash bottle is used whenever a fine, directed stream of distilled water is needed, as when rinsing down the sides of a glass vessel to ensure that no droplets of sample solution are lost.

Fig. 1.1 (a) Wash bottle, (b) desiccator.

(a) (b)

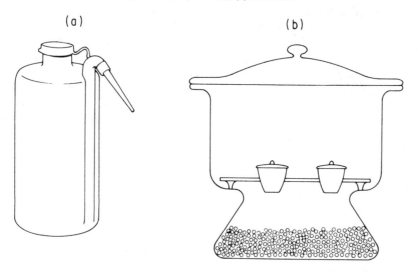

Stirring Rods

As the name implies, stirring rods are used for stirring solutions or suspensions, generally in beakers. The rods are cut from a length of solid glass rod, generally 3 or 4 mm in diameter, so as to extend about 6 or 8 cm from the top of the beaker. The ends should be fire-polished. In addition to their stirring function, stirring rods have other useful purposes. For example, they are used in transferring solutions from one vessel into another. When an aqueous solution is poured from the lip of a vessel such as a beaker, there is a tendency for some of the liquid to run down the outside surface of the glass. This is prevented by pouring the solution down a stirring rod, the rod being held in contact with the lip of the vessel and directing the flow of liquid into the receptacle (see Fig. 1.8). Stirring rods also serve as handles for "rubber policemen" (sections of rubber tubing sealed together at one end, with the other end slipped over a stirring rod, used to salvage small quantities of precipitates from the walls of beakers).

Desiccator

A desiccator is a vessel, usually of glass but occasionally of metal, which is used to equilibrate objects with a controlled atmosphere. Since the desiccator usually stands in the open, the temperature of this atmosphere generally approaches room temperature. It is normally the humidity of this atmosphere which is of interest. Objects such as weighing bottles or crucibles, and chemical substances, tend to pick up moisture from the air. The desiccator provides an opportunity for such materials to come to equilibrium with an atmosphere of low and controlled moisture content so that errors due to the weighing of water along with the objects can be avoided. A common type of desiccator is shown in Fig. 1.1(b).

The nature of the drying agent placed in the bottom of the desiccator determines the equilibrium partial pressure of water vapor in the desiccator space. Table 1.3 contains the results of Trusell and Diehl,[3] who studied the efficiency of various chemical desiccants. These workers studied 21 common drying agents by passing known volumes of wet nitrogen over the material, condensing the residual water in a liquid nitrogen trap, and weighing the water. The most powerful desiccant is not necessarily the best for a given application. Phosphorus pentoxide, for example, has a tendency to acquire a surface glaze as it picks up water, which prevents the bulk of the material from being effective. Calcium chloride, while rather poor with regard to equilibrium water vapor pressure, is inexpensive, has a fairly high capacity,[4] and is adequate for much analytical work.

---

[3] F. Trusell and H. Diehl, *Anal. Chem.*, **35**, 674 (1963).

[4] Capacity must be distinguished from equilibrium vapor pressure in describing desiccants. A desiccant may have a high capacity (i.e., it may be able to pick up a large weight of water vapor per unit weight of desiccant), but still leave much moisture in the air at equilibrium.

TABLE 1.3 EFFICIENCY OF CHEMICAL DESICCANTS

Material	Residual Water (μg/liter)†
$Mg(ClO_4)_2 \cdot 0.12H_2O$	0.2
$Mg(ClO_4)_2 \cdot 1.48H_2O$	1.5
BaO (96.2%)	2.8
Al_2O_3 (anhydrous)	2.9
P_2O_5	3.5
Molecular Sieve 5A (Linde)	3.9
$Mg(ClO_4)_2$ (88%) + $KMnO_4$ (0.86%)	4.4
$LiClO_4$ (anhydrous)	13
$CaCl_2 \cdot 0.18H_2O$	67
$CaSO_4 \cdot 0.02H_2O$ (Drierite)	67
Silica gel	70
NaOH (91%, Ascarite)	93
$CaCl_2$ (anhydrous)	137
$CaSO_4 \cdot 0.21H_2O$ (Anhydrocel)	207
$NaOH \cdot 0.03H_2O$	513
$Ba(ClO_4)_2$ (anhydrous)	599
CaO	656

†1 μg = 10^{-6} g.

Source: Data of F. Trusell and H. Diehl, *Anal. Chem.*, **35**, 674 (1963).

After reagents or objects such as crucibles have been dried in the oven, or perhaps at even higher temperatures, they are usually cooled to room temperature in the desiccator prior to weighing. When a hot object cools in the desiccator, a partial vacuum is created, and care must be taken in opening the vessel lest a sudden rush of air blow material out of a crucible or disturb the desiccant itself. For this reason, and also because glass is a very poor conductor of heat, it is usually best to allow a very hot object to cool well toward room temperature before it is placed in the desiccator. After a hot object has been placed in the desiccator, it is well to cover the vessel in such a way as to leave a small opening at one side. This allows air displaced by the warm object to reenter as the object cools, and hence minimizes the tendency to form a vacuum. The desiccator is completely closed during the final stages of cooling.

The desiccator cover should slide smoothly on its ground-glass surface. This surface should be lightly greased with a light lubricant such as Vaseline (never stopcock grease!). Needless to say, the desiccator should be scrupulously clean and should never contain exhausted desiccant. After filling the desiccant chamber, beware of dust from the desiccant in the upper part of the desiccator.

Pipets

Some common types of pipets are shown in Fig. 1.2. The *transfer pipet* is used to transfer an accurately known volume of solution from one container to another. The pipet should be cleaned if distilled water does not drain uniformly,

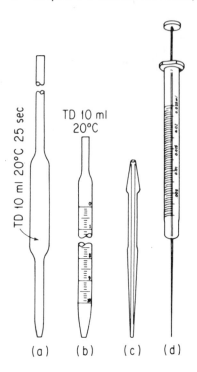

Fig. 1.2 Pipets: (a) transfer pipet, (b) measuring pipet, (c) lambda pipet, and (d) microliter syringe.

but leaves droplets of water adhering to the inner surface. Cleaning can be done with a warm solution of detergent or with cleaning solution (consult instructor).

The pipet is filled by gentle suction to about 2 cm above the etch line [Fig. 1.3(a)], using an aspirator bulb. Alternatively, a water aspirator can be used to apply suction. A long rubber tube is attached from the top of the pipet to the trap shown in Fig. 1.11. The tip of the pipet should be kept well below the surface of the liquid during the filling operation. The forefinger is then quickly placed over the top of the pipet [Fig. 1.3(b)], and the solution is allowed to drain out until the bottom of the meniscus coincides with the etched line. Any hanging droplets of solution are removed by touching the tip of the pipet to the side of the beaker, and the stem is wiped with a piece of tissue paper to remove drops of solution from the outside surface. The contents of the pipet are then allowed to run into the desired container, with care being taken to avoid spattering. With the pipet in a vertical position, allow the solution to drain down the inner wall for about 30 s after emptying, and then touch the tip of the pipet to the inner side of the receiving vessel at the liquid surface. A small volume of solution will remain in the tip of the pipet, but the pipet has been calibrated to take this into account; thus this small final quantity of solution is *not* to be blown out or otherwise disturbed. Pipets with damaged tips are not to be trusted.

Measuring pipets are graduated much like burets and are used for measuring volumes of solutions more accurately than could be done with graduated cylinders. However, measuring pipets are not ordinarily used where high accuracy is required.

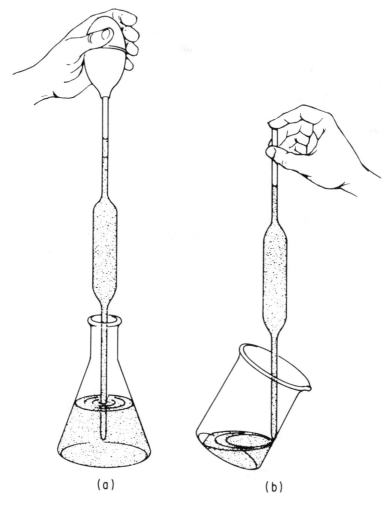

Fig. 1.3 (a) Filling pipet—liquid drawn above graduation mark, and (b) use of forefinger to adjust liquid level in pipet.

Two types of micropipets are shown in Fig. 1.2. The so-called *lambda* pipets are available in capacities of 0.001 to 2 ml, where 0.001 ml = 1 lambda. They are filled and emptied using a syringe. Those calibrated to *contain* a certain volume are rinsed with a suitable solvent. Those calibrated to *deliver* are not rinsed, but the last drop is forced out of the pipet with the syringe. Microliter syringes are widely used for delivering small volumes in such operations as gas chromatography. They can be bought equipped with stainless steel tips for use in injecting a sample into a closed system. The syringe shown in Fig. 1.2 has a capacity of 0.025 ml (25 μl, or 25 lambdas) and the smallest divisions correspond to 0.0005 ml. "Pushbutton" pipets, which make the transfer of liquids rapid and easy, are now available. Such a pipet consists of a syringe with a piston which can be operated by pressing a button at the top. Liquid is drawn into a disposable plastic tip and is then delivered by reversing the direction of the piston. Tips that deliver volumes of 0.001 to 1 ml (1 to 1000 μl) are available.

The National Institute of Standards and Technology specifies 20°C as the standard temperature for calibration of volumetric glassware. The use of such glassware at other temperatures leads to errors. However, the errors are normally small, and pipets can be used at "room temperature" without special precautions except in work of highest accuracy.

Burets

A common form of buret is shown in Fig. 1.4(a). The buret is used to deliver accurately known but variable volumes, mostly in titrations. The stopcock plug is made of either glass or Teflon. The Teflon stopcock requires no lubrication, but the glass plug should be lightly greased with stopcock grease (not one containing

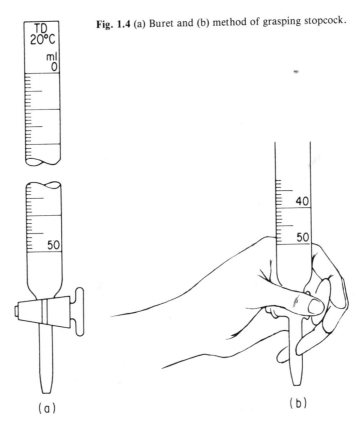

Fig. 1.4 (a) Buret and (b) method of grasping stopcock.

(a) (b)

silicones). If too heavy a coating is applied the stopcock may leak and also some of the grease may plug the buret tip. To lubricate a stopcock, remove the plug and wipe old grease away from both plug and barrel with a cloth or paper tissue. Make sure the small openings are not plugged with grease (pipe cleaners are helpful in this event). Then spread a thin, uniform layer of stopcock grease over the plug, keeping the application especially thin in the region near the hole in the plug. Finally, insert the plug in the barrel and rotate it rapidly in place, applying a slight inward pressure. The lubricant should appear uniform and transparent, and no particles of grease should appear in the bore.

Burets must be cleaned carefully to assure a uniform drainage of solutions down the inner surfaces. A hot, dilute detergent solution may be used for this purpose, especially if used in conjunction with a long-handled buret brush. Cleaning solution may also be used, applied hot for a few minutes or overnight at room temperature. The instructor should be consulted for directions on the proper use of cleaning solution. When not in use, the buret should be filled with distilled water and capped (paper cups or small beakers are convenient) to prevent the entry of dust.

It is poor practice to leave solutions standing in burets for long periods. After each laboratory period, solutions in burets should be discarded, and the burets rinsed with distilled water and stored as suggested above. It is especially important that alkaline solutions not stand in burets for more than short periods of time. Such solutions, which attack glass, cause stopcocks to "freeze," and the burets may be ruined.

The beginner must be very cautious in reading burets. In order to become familiar with the graduations and adept at estimating between them, much practice is needed early in the laboratory work. An ordinary 50-ml buret is graduated in 0.1-ml intervals and should be read to the nearest hundredth of a milliliter. An aqueous solution in a buret (or any tube) forms a concave surface referred to as a *meniscus*. In the case of solutions that are not deeply colored, the position of the bottom of the meniscus is ordinarily read (the top is taken if the solution is so intensely colored that the bottom cannot be seen, e.g., with permanganate solutions). It is most helpful to cast a shadow on the bottom of the meniscus by means of a darkened area on a paper or card held just behind the buret with the dark area slightly below the meniscus. (See Fig. 1.5.) Great care must be taken to avoid parallax errors in reading burets: the eye must be on the same level with the meniscus. If the meniscus is near a graduation that extends well around the buret, the correct eye-level can be found by seeking a position so that the graduation mark seen at the back of the buret merges with the same line at the front. A loop of paper encircling the buret just below the meniscus serves the same purpose.

Before a titration is started, it must be ascertained that there are no air bubbles in the tip of the buret. Such bubbles register in the graduated portion of the buret as liquid delivered if they escape from the tip during a titration, and hence cause errors. When a solution is delivered rapidly from a buret, the liquid running down

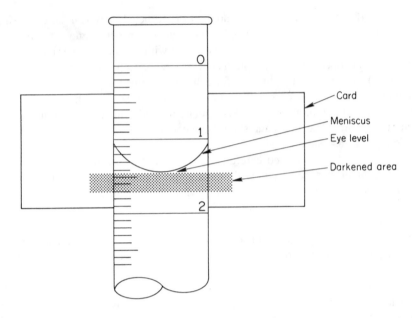

Fig. 1.5 Card for help in reading a buret. The eye should be level with the bottom of the meniscus. Here the reading is 1.42 ml.

the inner wall is somewhat detained. After the stopcock has been closed, it is important to wait a few seconds for this "drainage" before taking a reading.

In performing titrations, the student should develop a technique that permits both speed and accuracy. The solution being titrated, generally in an Erlenmeyer flask, should be gently swirled as the titrant is delivered. One way to accomplish this, while retaining control of the stopcock and permitting ease of reading the buret, is to face the buret, with the stopcock on the right, and operate the stopcock with the left hand from behind the buret while swirling the solution with the right hand [Fig. 1.4(b)]. The thumb and forefinger are wrapped around the handle to turn the stopcock, and inward pressure is applied to keep the stopcock seated in the barrel. The last two fingers push against the tip of the buret to absorb the inward pressure.

Volumetric Flasks

A typical volumetric flask is shown in Fig. 1.6. The flask contains the stated volume when filled so that the bottom of the meniscus coincides with the etched line. If the solution is poured from the flask, the volume delivered is somewhat less than the stated volume, and volumetric flasks are never used for measuring out solutions into other containers. They are used whenever it is desired to make a solution up to an accurately known volume.

When solutions are made up in volumetric flasks, it is important that they be

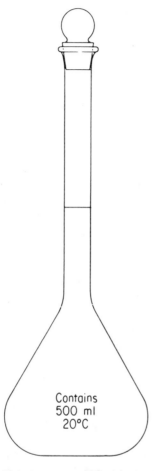

Contains
500 ml
20°C

Fig. 1.6 Volumetric flask.

well mixed. This is accomplished by repeatedly inverting and shaking the flask. Some analysts make a practice of mixing the solution thoroughly before the final volume has been adjusted, and mixing again after the flask has been filled to the mark: it is easier to agitate the solution vigorously when the narrow upper portion of the flask has not been filled.

Solutions should not be heated in volumetric flasks, even those made of Pyrex glass. There is a possibility that the flask may not return to its exact original volume upon cooling.

Most volumetric flasks have ground-glass or polyethylene stoppers, screw caps or plastic snap caps. Alkaline solutions cause ground-glass stoppers to "freeze" and thus should never be stored in flasks equipped with such stoppers.

When a solid is dissolved in a volumetric flask, the final volume adjustment should not be made until all the solid has dissolved. In certain cases marked volume changes accompany the solution of solids, and these should be allowed to take place before the volume adjustment is made.

Funnels and Filter Paper

In gravimetric procedures the desired constituent is often separated in the form of a precipitate. This precipitate must be collected, washed free of undesirable contaminants from the mother liquor, dried, and weighed, either as such or after conversion into another form. Filtration is the common way of collecting precipitates, and washing is often accomplished during the same operation. Filtration is carried out with either funnels and filter paper or filtering crucibles. The important factors in choosing between the two are the temperature to which the precipitate must be heated to convert it into the desired weighing form and the ease with which the precipitate may be reduced.

The cellulose fibers of filter paper have a pronounced tendency to retain moisture, and a filter paper containing a precipitate cannot be dried and weighed as such with adequate accuracy. It is necessary to burn off the paper at a high temperature. During the burning, reducing conditions due to carbon and carbon monoxide prevail in the vicinity of the precipitate. Thus precipitates that cannot be heated to high temperatures or that are sensitive to reduction are normally not filtered using filter paper; filtering crucibles of the types described in a later section are employed. Some of the techniques given here, however, will apply to all types of filtration.

Various types of filter paper are available. For quantitative work, only paper of the so-called "ashless" quality should be used. This paper has been treated with hydrochloric and hydrofluoric acids during its manufacture. Thus it is low in inorganic material and leaves only a very small weight of ash when it is burned. (A typical figure for the ash from one circular paper 11 cm in diameter is 0.13 mg.) The weight of ash is normally ignored; for very accurate work, a correction can be applied, since the weight of ash is fairly constant for the papers in a given batch.

Within the ashless group, there are further varieties of paper that differ in porosity. The nature of the precipitate to be collected dictates the choice of paper. "Fast" papers are used for gelatinous, flocculent precipitates such as hydrous iron (III) oxide and for coarsely crystalline precipitates such as magnesium ammonium phosphate. Many precipitates that consist of small crystals (e.g., barium sulfate), will pass through the "fast" papers. "Medium" papers require a longer time for filtration, but retain smaller particles and are the most widely used. For very fine precipitates such as silica, "slow" paper is employed. Filtration at best is rather time consuming, and the analyst should use the fastest paper consistent with retention of the precipitate.

Filter paper is normally folded so as to provide a space between the paper and the funnel, except at the top of the paper, which should fit snugly to the glass. The procedure is shown in Fig. 1.7. The second fold is made so that the ends fail to match by about $\frac{1}{8}$ in. Then the paper is opened into a cone. The corner of the outside fold on the thicker side is torn off in order to fit the paper to the funnel more easily and to break up a possible air passage down the fold next to the funnel.

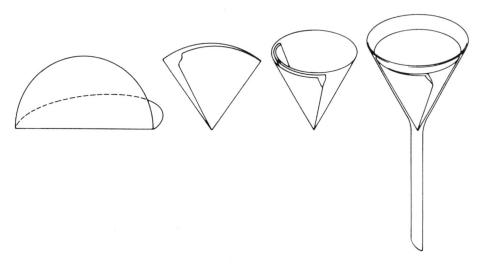

Fig. 1.7 Folding filter paper.

With the paper cone held in place in the funnel, distilled water is poured in. A clean finger, applied cautiously to prevent tearing the fragile wet paper is used to smooth the paper and obtain a tight seal of paper to glass at the top. Air does not enter the liquid channel with a properly fitted paper, and thus the drainage from the stem of the funnel establishes a gentle suction which facilitates filtration. A malfunctioning filter can seriously delay an analysis; it is preferable by far to reject such a filter and prepare a new one.

Filter paper circles are available in various diameters. The size to be used depends upon the quantity of precipitate, not the volume of solution to be filtered. Larger paper than necessary should be avoided; the paper and the funnel should match with regard to size. It is especially important that the paper not extend above the edge of the glass funnel, but come within 1 or 2 cm of this edge. The precipitate should occupy about one-third of the paper cone and never more than one-half.

Technique of Washing and Filtering a Precipitate

Usually a precipitate is washed, either with water or a specified wash solution, before it is dried and weighed. The washing is generally carried out in conjunction with the filtration step (Fig. 1.8), wherein the precipitate is separated from its mother liquor in compact form. Once the precipitate is in the filter, it can be washed by passing wash solution through the filter. However, this technique is often rather inefficient; the wash solution does not penetrate uniformly into the compact mass of precipitate. It is usually preferable to wash the precipitate by decantation, at least in cases where the precipitate settles rapidly from suspension. The supernatant mother liquor is carefully poured off through the filter while as much of the precipitate as possible is retained in the beaker. The precipitate is then stirred with

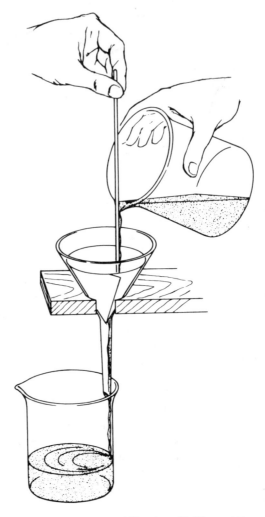

Fig. 1.8 Technique of filtration with filter paper.

wash solution in the beaker, and the washings are decanted through the filter. This washing is repeated as often as desired,[5] until, in the final instance, the precipitate is not allowed to settle but is poured into the filter along with the wash solution. Residues of precipitate remaining in the beaker are usually transferred to the filter by a directed jet from a wash bottle, as shown in Fig. 1.9. If the precipitate tends to adhere to the glass, the last traces may be scavenged by means of a rubber policeman. The precipitate is then wiped from the policeman with a small bit of filter paper, which is added to the paper in the funnel for ignition.

[5] It should be noted that several washings with small volumes of solution are more effective than one washing with the same total volume of wash solution.

The stem of the funnel should extend well into the vessel receiving the filtrate, and the tip of the stem should touch the inner surface of the vessel to prevent spattering of the filtrate. All transfers into the funnel should be made with the aid of a stirring rod, and care must be taken that no drops of solution are lost. The filtrate should be examined for turbidity; sometimes small amounts of precipitate run through the filter early in the filtration, but can be caught by refiltering through the same filter after its pores have been somewhat clogged by collected precipitate.

Ignition of Precipitate with Filter Paper

After the filter paper has drained as much as possible in the funnel, the top of the paper is folded over to encase the precipitate completely. Using great care to avoid tearing the fragile, wet paper, the bundle of paper is transferred from the funnel to the prepared crucible (see the discussion of crucibles below). It is better to handle the paper where it is three layers thick rather than by the other side. The next steps in the ignition of the material in the crucible are generally as follows:

Fig. 1.9 Use of wash bottle in transferring precipitate.

1. Drying the Paper and Precipitate. This may be done in an oven at temperatures of 100 to 125°C if the schedule permits setting aside the experiment at this stage. If the ignition is to be followed through immediately, the drying may be accomplished with a burner. Place the covered crucible in a slanted position in a clay or silica triangle, and place a small flame beneath the crucible at about the middle of the underside. Too strong heating must be avoided; the flame should not touch the crucible, and the drying should be leisurely.

2. Charring the Paper. (See Fig. 1.10.) After the precipitate and paper are entirely dry, the crucible cover is set ajar to permit access of air, and the heating is increased to char the paper. Increase the size of the flame slightly and move it back under the base of the crucible. The paper should smoulder, but must not burn off with a flame. If the paper bursts into flame, cover the crucible immediately to extinguish it. Small particles of precipitate may be swept from the crucible by the violent activity of escaping gases; also, under these conditions, carbon of the paper may reduce certain precipitates which can be safely handled in filter paper under less vigorous conditions. Care must be taken that reducing gases from the flame are not deflected into the crucible by the underside of the cover. During the charring, tarry organic material distills from the paper, collecting on the crucible cover. This is burned off later at a higher temperature.

Fig. 1.10 Ignition of precipitate.

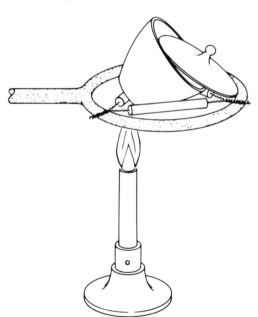

3. Burning Off the Carbon from the Paper. After the paper has completely charred and the danger of its catching fire is past, the size of the flame is increased until the bottom of the crucible becomes red. This should be done gradually. The carbon residue and the organic tars are burned away during this stage of the ignition. The heating is continued until this burning is complete, as evidenced by the disappearance of dark-colored material. It is well to turn the crucible from time to time so that all portions are heated thoroughly. Sometimes it is necessary to direct special attention to the underside of the cover to remove the tarry material collected there.

4. Final Stage of Ignition. To conclude the ignition, place the crucible upright, removing the cover to admit air freely, and heat at the temperature recommended for the particular precipitate. A Tirrill burner will heat a covered porcelain crucible to about 700°C, and a Meker burner will give a temperature roughly 100°C higher. With platinum crucibles, temperatures about 400°C higher can be obtained. The ignition is continued until the crucible has reached constant weight; that is, until the difference between two weighings with a heating period in between is less than about 0.5 mg.

Filtering Crucibles

Certain precipitates cannot be handled with filter paper, either because they are too easily reduced or because they cannot be heated to a temperature adequate to burn off the paper. Such precipitates are filtered by means of *filtering crucibles,* several types of which are pictured in Fig. 1.11. The *Gooch,* which is seldom used today, is a porcelain crucible with a number of small holes in the bottom. A mat of asbestos is formed on the inside of the bottom when the crucible is prepared for use, and this mat is the filtering medium. It is prepared by forming a suspension of specially cut and treated asbestos fibers in water, and pouring the suspension through the crucible under suction. Gooch crucibles can be ignited at high temperatures and are adequate for many different precipitates.

For precipitates that need not be heated above 500°C or so, *sintered* or *fritted glass crucibles* may be used. These crucibles are made of glass, and have a bottom of sintered ground glass fused to the body of the crucible. They are mounted in a suitable holder (as shown in Fig. 1.11) and suction is applied. Sintered glass crucibles are available in varying porosities for handling various types of precipitates. In the Corning Glass Co. Pyrex line, the three porosities designated coarse, medium, and fine (the letters C, M, and F appear near the upper edge of the glass crucible) serve most analytical usages. Although temperatures up to 500°C are said to be safe, sintered glass crucibles must be heated very gradually if such a temperature is to be approached without damage. It must be kept in mind that strong alkalies attack the crucibles, especially the sintered filtering disks. The crucibles are cleaned with solvents appropriate to the particular contamination at hand. For

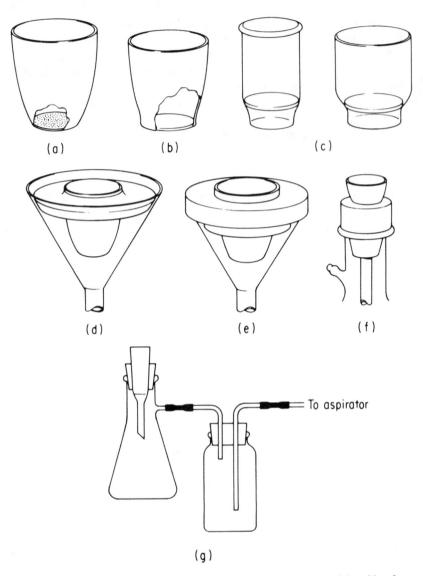

Fig. 1.11 Filtering crucibles: (a) Gooch; (b) porous porcelain; (c) fritted glass, high and low forms; (d) Sargent holder; (e) Bailey holder; (f) Walter holder; and (g) filtration with suction, Walter holder.

three reasons sintered glass crucibles are never heated directly over a flame. First, they may be broken; second, carbon, which is very difficult to remove, may be deposited in the fritted disk; and third, reducing gases from the flame may penetrate to the precipitate via the porous bottom. If a burner must be used to attain the desired temperature, the sintered glass crucible is placed within an ordinary porcelain crucible.

For precipitates which must be ignited at very high temperatures, *porous porcelain crucibles* may be employed. These are porcelain crucibles with unglazed, porous bottoms through which precipitates can be filtered off under suction. As with sintered glass crucibles, the porous porcelain crucibles must not be heated with an open flame. They are ignited within an ordinary porcelain crucible unless a furnace is used. The porous bottom is readily attacked by a strong alkali.

Crucibles for Ignition of Precipitates in Filter Paper

Different kinds of crucibles have been suggested for one purpose or another, but we may confine our discussion to the two most widely used, *porcelain* and *platinum.* Platinum is usually preferred, although porcelain, which is inexpensive but adequate for many purposes, is also widely used. There are certain cases where platinum must not be used (see below).

Porcelain crucibles are resistant to attack by many reagents; however, they are attacked by some, notably alkaline substances and hydrofluoric acid. They can be heated to high temperatures (about 1200°C), and their weight changes very slightly with strong and prolonged heating. It must be remembered that, although they can be heated safely to about 1200°C, porcelain crucibles do not attain this temperature over a burner. A platinum crucible over the same burner reaches a much higher temperature. In a furnace, on the other hand, porcelain and platinum come to the same temperature. As an example of the effect of this temperature difference between porcelain and platinum when a burner is used, note that in porcelain calcium oxalate cannot be ignited satisfactorily to the oxide, while in platinum the conversion is completed readily.

For many purposes, platinum crucibles are superior to porcelain. Platinum crucibles reach a higher temperature over a burner (but not in a furnace), and they cool much more rapidly because of the high thermal conductivity of the metal. Platinum is a very inert metal, and it is resistant to certain reagents that attack porcelain. A notable example of this is the fact that a precipitate can be treated with hydrofluoric acid in a platinum crucible. This is used, for example, in the accurate determination of silica by treating the impure SiO_2 precipitate with hydrofluoric acid, volatilizing silicon tetrafluoride, and obtaining the pure silica by difference.

On the other hand, platinum is attacked by certain substances, and because of its great expense, care must be exercised in its use. Your instructor will caution you as necessary on the use of platinum.[6]

The idea of *constant weight* should be explained more thoroughly. The weight of a precipitate in a crucible is normally obtained by difference: first the empty

---

[6]If platinumware is employed, a detailed discussion on the proper care of platinum should be studied. A good discussion is given in "Notes on the Care and Use of Platinum Crucibles and Dishes," a pamphlet circulated by Matthey Bishop, Inc., Malvern, PA 19355.

crucible is weighed, and then the crucible plus ignited precipitate. The ignition of the precipitate sometimes involves a chemical conversion—for example,

$$CaC_2O_4 \cdot 2H_2O \longrightarrow CaO \quad \text{or} \quad MgNH_4PO_4 \cdot 6H_2O \longrightarrow Mg_2P_2O_7$$

—into the final weighing form, or sometimes merely drying. This change can be assumed to be complete when the weight of the precipitate no longer changes with further heating. The precipitate is heated at the desired temperature for a reasonable length of time, and then cooled and weighed. It is then reheated, usually for a shorter period of time, and weighed again. If the two weighings agree (within, say, 0.2 to 0.5 mg) the ignition is considered complete. The precipitate has been ignited to "constant weight." Now, since the weight of precipitate is obtained by difference, the weight of the crucible alone must also be accurately known. Further, it is sometimes found that the weight of a crucible varies somewhat with the conditions under which it was prepared and dried. Thus the empty crucible is usually ignited *exactly* as the crucible plus precipitate will be ignited later. This ignition is continued until the crucible itself is at constant weight.

Special Instruments

Not too many years ago, the analytical chemist used only the apparatus described above, plus the analytical balance, for almost all determinations. A striking change has taken place, and nowadays pH meters, spectrophotometers, polarographs, gas and liquid chromatographs, and other complex instruments are found in most analytical laboratories. Directions for the use and care of instruments are necessarily specialized, and are best obtained from the manufacturer's bulletins and from personal instruction by experienced people. We include here only a few general remarks to fill out our discussion of analytical apparatus.

The rule of greatest importance is that no instrument should ever be touched by a person unfamiliar with the directions for its proper use and the precautions against damaging it. Some instruments contain fragile components which may be injured by improper handling, and sometimes a carefully worked out calibration may be ruined by manipulation of the wrong knobs.

The other rule that must always be remembered is that an instrument should never be used by a person who has not thought through its advantages and limitations for the job at hand, who does not have a proper estimate of the reliability of the data obtained, and who cannot interpret correctly the significance of the instrumental measurement and apply it with intelligence. Meaningless measurements are made every day by imposters masquerading as chemists. Anyone can learn to turn knobs and read galvanometers, but the assurance that a measurement has been made on the best possible system must come from a well-trained chemist.

TABLE 1.4 TOLERANCES FOR VOLUMETRIC GLASSWARE (ml)

Capacity, Less Than and Including:	Volumetric Flasks	Transfer Pipets	Burets
2		0.006	
5		0.01	0.01
10		0.02	0.02
25	0.03	0.03	0.03
50	0.05	0.05	0.05
100	0.08	0.08	0.10
200	0.10	0.10	
500	0.15		
1000	0.30		

CALIBRATION OF VOLUMETRIC GLASSWARE

Introduction

Table 1.4 shows some of the tolerance values established for volumetric glassware by the National Institute of Standards and Technology.[7] It may be noted that glassware meeting these specifications is adequate for all but the most exacting work of the analytical laboratory. Such glassware, stated by the manufacturer to conform to NIST standards, may be purchased.[8] For an extra fee, it is also possible to obtain glassware that has actually been tested by the NIST. Because of the expense, however, the beginning analytical laboratory will rarely be equipped with glassware guaranteed to meet NIST tolerance specifications. Less expensive glassware stated by the manufacturer to meet tolerances about double those of the NIST is available,[9] but even this glassware may not be furnished the student. For this reason, and also because the history of a particular item involving possible damage may be obscure, it is often advisable for students to calibrate their volumetric glassware. Of course, the individual instructor will make the decision in any case.

Since most analytical work involves dilute, aqueous solutions, water is generally used as the reference material in the calibration of volumetric glassware. The general principle in calibration is to determine the weight of water contained in or delivered by a particular piece of glassware. Then, with the density of water known, the correct volume is found.

The units of volume commonly employed in analytical chemistry are the *liter* and the *milliliter*. The liter was formerly defined as the volume occupied by 1 kg

---

[7] Nat. Bur. Standards Circ. 434 (1941).

[8] For example, in the Kimball Glass Co. Kimax Class A line and the Corning Glass Works Pyrex line of volumetric glassware.

[9] Kimball Glass Co. Kimax volumetric glassware.

of water at the temperature of its maximum density (about 4°C) under a pressure of 1 atm. In 1964 the Twelfth General Conference on Weights and Measurements, meeting in Paris, France, abolished this definition and instead made the liter a special name for the cubic decimeter. This new definition eliminates the previous discrepancy of 28 parts per million between the milliliter and cubic centimeter (1 ml was 1.000028 cc), and these two units are now equivalent.

The National Institute of Standards and Technology has specified 20°C as the temperature at which glassware is calibrated. Since the laboratory temperature will usually not be exactly 20°C, glassware must, strictly speaking, be corrected when used at other temperatures because of errors due to expansion (or contraction) of both the glass vessel itself and the solution contained therein. The coefficient of expansion of glass is sufficiently small that the correction required for this factor is negligible for most work (it amounts to the order of 1 part per 10,000 for a change of 5°C). The change in the volume of the solution itself, on the other hand, is more important, but it can still be ignored in many cases if the working temperature is not far removed from 20°C (the volume change is of the order of 1 part per thousand over a 5°C range).

As noted above, calibration data are secured by converting weight of water into volume via the density. Tables showing the density of water at various temperatures are available in handbooks. However, the data in such tables are usually given on the basis of weights in vacuo, while the actual weighings are made in air. Since the water being weighed generally displaces more air than do the weights, it is necessary to correct the weighings for the buoyancy effect of the air if such handbook tables are to be used (see Chapter 2 for a fuller discussion of this buoyancy effect). On the other hand, we may change in vacuo densities into densities that would be obtained in air with steel weights, and use these directly with our weighings obtained in air. The values in Table 1.5, which are the reciprocals of such adjusted densities, may be used directly by the student without taking the buoyancy effect into consideration.

Methods of Calibration Commonly Employed

There are three general approaches to the calibration of volumetric glassware that are in wide use and with which the student should be familiar.

1. The first method, which we may designate as a *direct, absolute calibration,* is based on the principles outlined above. The volume of water delivered by a buret or pipet, or contained in a volumetric flask, is obtained directly from the weight of the water and its density. Directions are given below for the calibration of a buret, a pipet, and a volumetric flask using this method.
2. Volumetric glassware is sometimes calibrated by comparison with another vessel previously calibrated directly. We may refer to this as an *indirect, absolute calibration, or calibration by comparison.* This method is especially convenient if many pieces of glassware are to be calibrated, and

TABLE 1.5 VOLUME OF 1 G OF WATER WEIGHED IN AIR WITH STEEL WEIGHTS AT VARIOUS TEMPERATURES

°C	ml	°C	ml
10	1.0013	21	1.0030
11	1.0014	22	1.0033
12	1.0015	23	1.0035
13	1.0016	24	1.0037
14	1.0018	25	1.0040
15	1.0019	26	1.0043
16	1.0021	27	1.0045
17	1.0022	28	1.0048
18	1.0024	29	1.0051
19	1.0026	30	1.0054
20	1.0028		

it is sufficiently accurate for all ordinary usages provided that the reference vessel itself has been accurately calibrated. Calibrating bulbs are not available in many student laboratories, and specific directions for their use are not included in this chapter. They are not difficult to use if proper care is taken. Students may obtain directions from their instructors if they are to use such equipment.

3. Sometimes it is necessary to know only the relationship between two items of glassware without knowing the absolute volume of either one. This situation arises, for example, in taking an aliquot portion of a solution. Suppose that it is desired to titrate one-fifth of an unknown sample. The unknown might be dissolved, appropriately treated preparatory to the titration, and diluted to volume in a 250-ml volumetric flask. A 50-ml pipet would then be used to withdraw an aliquot for titration. For the calculations in this analysis, it would not be necessary to know the exact volume of the flask or the pipet, but it would be required that the pipet hold exactly one-fifth as much solution as the flask. The method used for a *relative calibration* of this sort simply involves discharging the pipet five times into the flask and marking the level of the meniscus on the flask.

Calibration of Buret

The buret should be thoroughly cleaned so that it drains well, and the stopcock should be properly lubricated. Fill the buret with water and test for leakage, making a reading to the nearest 0.01 ml and repeating the reading after waiting at least 5 min. No noticeable change should have occurred. During the waiting period, weigh a stoppered 125-ml Erlenmeyer flask (or other suitable container) to the nearest milligram. Record this weight.

Fill the buret with distilled water that is at the temperature of the laboratory. This temperature must be measured and recorded. Then sweep any air bubbles from the tip of the buret by opening the stopcock to permit rapid outflow. Now withdraw water more slowly until the meniscus is at, or slightly below, the zero mark on the buret. After drainage is complete (at least 30 s), read the buret to the nearest 0.01 ml. Record this "initial" reading. Remove any hanging drop of water from the tip of the buret by touching it lightly to the side of a vessel such as a beaker. Now run about 10 ml of water from the buret into the weighed flask. The tip of the buret should extend somewhat into the mouth of the flask, and care should be taken against spattering. The neck of the flask where contact with the stopper is made should not be wet. Quickly stopper the flask. Read the buret after allowing time for drainage and record this "final" reading. Then weigh the stoppered flask to the nearest milligram. Record this weight. Tabulation of the data as suggested in Table 1.6 is convenient.

Now refill the buret and obtain another "initial" reading. Run about 20 ml of water into the flask, obtain the "final" buret reading, and reweigh the flask. Note that we are calibrating the buret in 10-ml intervals, but starting each time from an initial reading near zero since in titrations we generally start near this point.

This process is repeated for the 30-, 40-, and 50-ml volumes. The flask, of course, should never be allowed to exceed the capacity of the balance. It should first be emptied, the neck dried with a clean towel, and reweighed.

For each calibration interval, multiply the weight of water by the appropriate value from Table 1.5 to obtain the actual volume of the water. The difference between this actual volume and the apparent volume (from the buret readings) is, of course, the correction. If the actual volume is larger than the apparent volume, the correction obtained by subtracting in this way will be positive, which means that it is to be added to the buret reading in future work with the buret.

The calibration should be repeated as a check on the work. Duplicate results should agree within 0.04 ml. With the weighings carried out to the nearest milligram, as directed above, errors in weighing should not affect the results because the weight is actually needed only to the nearest 0.01 g. Hence failure of duplicate results to agree is generally due to (1) leakage around the buret stopcock, (2) failure to wait for drainage before reading the buret, or (3) careless technique in collecting or handling the container and stopper.

Since our calibration is performed over 10-ml intervals, students often ask about the correction to be applied to a buret reading of an intermediate value such

TABLE 1.6   SAMPLE DATA FOR CALIBRATION OF BURET

Initial buret reading	0.38	0.49
Final buret reading	10.16	20.16
Apparent volume, ml	9.78	19.67
Initial weight of flask	62.576	72.311
Final weight of flask	72.311	91.832
Weight of water, g	9.735	19.521
Temperature, °C	24	24
Actual volume, ml	9.77	19.59
Correction, ml	−0.01	−0.08

as 46 ml. Ordinarily, applying the correction of the nearest interval, in this case 50 ml, will be sufficient. It is perhaps somewhat better to interpolate between the known corrections. This is done most conveniently by a graphical method. The student may plot corrections against intervals 10, 20 ml, . . . , connecting the points with straight lines so as to obtain linear interpolations by simple inspection of the graph. Such an interpolation obviously is not likely to reflect the true situation within the buret, but it will usually be more than sufficiently accurate for the need at hand. The student must use good judgment in consideration of the precision desired in the buret reading. For example, an error of 0.04 ml in a 40-ml buret reading represents only 1 ppt.

Calibration of Pipet

The method is essentially the same as that described above for a buret. The pipet should naturally be carefully cleaned and rinsed before calibration. Weigh to the nearest milligram a clean, stoppered Erlenmeyer flask of adequate capacity. Now fill the pipet to a level above the etched line, using a beaker of distilled water which is at the temperature of the laboratory. This temperature must be measured, of course. Dry the outside of the pipet with a clean towel, and then release the finger pressure to allow the liquid level to fall to the etched line. The pipet should be held in a vertical position with the etched line at eye-level for this operation. Hold the liquid level so that the bottom of the meniscus coincides with the etched line, and touch the tip of the pipet to the side of the beaker to remove any hanging drop. Then discharge the contents of the pipet into the weighed container. It is best to insert the tip of the pipet well into the container and to keep the tip touching the inner wall of the container during the delivery. With the tip still touching the container, allow the pipet to drain for 20 to 30 s after the flow ceases. Do not disturb the last portion of water remaining in the tip. Then stopper the container and reweigh it. Calculate the volume of water delivered by the pipet from the weight and the appropriate value taken from Table 1.5. The calibration should be repeated, and the duplicate results should not differ by more than 1 ppt. Common errors are (1) warming the contents of the pipet by holding the enlarged portion in the palm of the hand, (2) failure to allow sufficient drainage time, (3) disturbing the residue of water that should remain in the tip of the pipet, and (4) general carelessness in handling the weighed container and stopper.

Calibration of Volumetric Flask

For large volumetric flasks, an ordinary analytical balance cannot be used. For example, a 250-ml flask may weigh as much as 100 g, and may weigh 350 g or so when filled with water. This, of course, exceeds the capacity of an analytical balance, and a large-capacity balance must be used. Because of possible inequality of the balance arms of such balances, weighing by the method of substitution, as described below, is recommended.

The flask should be thoroughly cleaned and rinsed, and then clamped in an inverted, vertical position until dry. Now stopper the flask and place it on the left-hand pan of the balance. Add copper shot to a beaker on the right-hand pan of the balance until the rest point of the balance comes on scale. Note the position of the rest point, and with the copper shot undisturbed, replace the flask with weights just appropriate to give the same rest point. Record this weight.

Now, using a small funnel, add distilled water at room temperature to the flask until the flask is nearly filled. Remove the funnel, being careful to avoid leaving drops of water on the neck of the flask above the mark. (If such drops do appear, remove them with a length of filter paper.) Very carefully complete the filling of the flask up to coincidence of the meniscus with the etched mark by means of a pipet or dropper. Reweigh the flask by substitution as above. From the weight of water, the temperature, and the appropriate value from Table 1.5, calculate the volume of water that the flask contains. The calibration should be checked, of course, by repeating the procedure. Duplicate results should agree within 0.1 ml for a 250-ml flask.

Relative Calibration of 50-ml Pipet and 250-ml Volumetric Flask

The pipet and flask should be carefully cleaned and rinsed. The flask should be dried. Using an approved technique, introduce into the flask five 50-ml portions of distilled water from the pipet. The tip of the pipet should extend well into the flask to avoid splashing, and the pipet should be operated carefully with regard to drainage and the other points noted previously in this chapter. Finally, mark the position of the bottom of the meniscus on the neck of the flask by means of the upper edge of a gummed label. The calibration should be repeated.

RECORDING LABORATORY DATA

There are three main requirements for the recording of data obtained in the analytical laboratory. These may be briefly expressed as follows: The record (1) should be complete, (2) should be intelligible to any reasonably competent chemist, and (3) should be easy to find on short notice. These requirements may be met by adherence to the following rules.

1. Students should have a bound notebook for recording their laboratory data, calculations, results, and all other matters pertinent to the analysis of a sample. The pages of the notebook should be numbered, and a table of contents should be developed so that any given experiment can be quickly found.
2. All data obtained in the laboratory should be recorded directly in the notebook at the time the work is performed. Especially forbidden is the recording of data on loose paper with the idea of copying it into the notebook later. While neatness may be sacrificed somewhat by taking the

TABLE 1.7 SAMPLE NOTEBOOK PAGE FOR SODA ASH ANALYSIS

Soda Ash Analysis

Method: sample dissolved in water and titrated with standard
HCl solution, using modified methyl orange indicator

Reaction: $Na_2CO_3 + 2HCl \rightarrow CO_2 + H_2O + 2NaCl$

	I	II	III
Wt. of sample	0.3276 g	0.3342 g	0.3128 g
Final buret reading, HCl	35.86 ml	36.32 ml	34.56 ml
Initial buret reading, HCl	0.18 ml	0.10 ml	0.38 ml
Volume HCl	35.68 ml	36.22 ml	34.18 ml
Initial buret reading, NaOH	0.37 ml	0.57 ml	0.92 ml
Final buret reading, NaOH	0.06 ml	0.37 ml	0.57 ml
Volume NaOH	0.31 ml	0.20 ml	0.35 ml
Volume relation of NaOH to HCl: 1.00 ml NaOH = 0.95 ml HCl			
ml HCl equiv. to NaOH	0.29 ml	0.19 ml	0.33 ml
Total HCl req. for sample	35.39 ml	36.03 ml	33.85 ml
Normality of HCl: 0.1076			
$\% \ Na_2CO_3 = \dfrac{\text{ml HCl} \times N \times EW}{\text{mg sample}} \times 100$	61.60%	61.49%	61.70%
Average $\% \ Na_2CO_3$		61.60%	
Average deviation, ppt		1.2 ppt	

notebook directly into the laboratory, the prevention of loss of data and errors in transcribing them more than counterbalances this.

3. Entries should be recorded in ink. If a mistake is made and a recorded value is invalidated, it is not to be erased, but is crossed out so as to be still legible. A notation as to why it is rejected is made in the notebook.

4. The data in the notebook should be organized and recorded in a systematic way. This benefits the student because it is then relatively easy to locate errors in the analytical calculations; the student may thus be saved repeating an entire determination in order to obtain satisfactory results. To facilitate an orderly presentation of the data, the student should plan how best to record it before the experiment is actually begun. It is especially helpful to arrange beforehand a table in which the experimental data, the calculations, and the final results can be entered systematically.

5. The rules regarding significant figures (see Chapter 2 of the text)[10] should be followed in recording data in the notebook.

Table 1.7 shows an example of a satisfactory laboratory record for a typical volumetric analysis. The opposite page in the notebook may be used for arithmetic

[10]R. A. Day, Jr., and A. L. Underwood, *Quantitative Analysis,* 6th ed., Prentice Hall, Inc., Englewood Cliffs, NJ, 1991 (the text which this manual is designed to accompany).

TABLE 1.8 SAMPLE REPORT CARD FOR SODA ASH UNKNOWN

10/18/91	Name		Soda Ash Unknown No. 186	
		Normality of HCl: 0.1076		
	Wt sample	ml HCl	% Na_2CO_3	Deviation
1	0.3276 g	35.39 ml	61.60	0.00
2	0.3342 g	36.03 ml	61.49	0.11
3	0.3128 g	33.85 ml	61.70	0.10

Average % Na_2CO_3: 61.60

Average deviation: 0.07 in 61.60 or 1.2 ppt

calculations and other material which does not fit well into the tabulation. The student should, of course, follow any style suggested by the individual instructor. Some instructors grade unknowns directly from the notebook, while others prefer that the student summarize results on a 3- by 5-in. card somewhat as shown in Table 1.8. Still others may ask the student to report results on a computer terminal.

There are few situations where graduate chemists (indeed, any professional) will not be expected to be neat, efficient, systematic, and able to locate quickly the results of their work and interpret them effectively for their superiors. The notebook in analytical chemistry should provide a good breeding ground for desirable habits.

SAFETY IN THE ANALYTICAL LABORATORY

It is unfortunate that laboratory safety is not usually emphasized sufficiently in college courses. Generally speaking, our colleagues in industry are much more safety conscious. It is largely up to the individual instructor to see that safety regulations are enforced, but a few general remarks are quite in order in this book.

Injuries in the laboratory are usually due to one of the following (although we realize, of course, that there is overlap among these categories in many accidents): (1) fire, (2) poisons, (3) broken glass, and (4) explosions. Fire is not a common danger in the beginning analytical laboratory because inflammable substances (organic solvents, for example) are not used extensively. Whenever a solvent such as ether or alcohol is used, care must be taken that no open flames are in the vicinity. Similarly, because of the nature of the work, the beginning analytical student is not very likely to experience an explosion. Great care must be exercised in dealing with substances or apparatus where an explosion may occur. Perchloric acid comes to mind as an example of a widely used substance that can be very dangerous if not used properly. The possibility of an explosion must always be considered whenever a container is at hand whose contents are under pressure, such as cylinders of compressed gases. Cuts from broken glass are common when glass tubing is improperly pushed into a tight hole in a rubber stopper and when unusual pressure is applied to a thin glass vessel such as a beaker. Where there is a possibility that necessary hand pressure will cause glass to break, the hand should be protected with a glove or towel.

Perhaps the greatest danger in the analytical laboratory is from poisons, if we include in this class corrosive substances such as strong acids and bases that readily attack human tissues. Such reagents should be handled with the greatest care. When a substance such as sulfuric acid is spilled on the skin or splashed into an eye, the severity of the resulting burn may depend upon the speed with which the situation is handled. There may be no time in which to look for expert help, and thus each student should know beforehand the emergency treatment to be undertaken. With acid or alkali burns, the first step, which is to be taken *immediately,* ignoring the common courtesies and forgetting any possible embarrassment, is to wash the affected area with copious quantities of cool water. This should be followed by washing with a solution of a weak base such as bicarbonate in the case of acid burns, or a dilute solution of a weak acid such as acetic for alkali burns. The instructor should see that each student knows exactly where they are located. Prevention is far superior to any treatment, of course, and proper caution will prevent most acid burns. Under no circumstances should corrosive solutions be sucked into a pipet by mouth. In most laboratories it is required that the eyes be protected by presciption glasses or special safety glasses at all times.

While it is tempting to be most impressed by strong acids and bases, it should be borne in mind that nearly all the chemicals encountered in the laboratory are poisons. "Familiarity breeds contempt," and we often forget that common substances like hydrogen sulfide, benzene, carbon tetrachloride, and the vapor from mercury can be fatal to human beings. We cannot discuss the toxicology of all these poisons individually, but we must warn the student, and especially the instructor, to be alert to the dangers involved in laboratory work.

Safety charts, posters, and signs for the laboratory may be obtained from a number of sources, such as (1) Manufacturing Chemists' Assoc., Inc., 1825 Connecticut Avenue, N.W., Washington, DC 20009, and (2) Fisher Scientific Co., 711 Forbes Avenue, Pittsburgh, PA 15219. A three-volume book entitled *Safety in the Chemical Laboratory,* by Norman V. Steere, is available from the Journal of Chemical Education, Office of Publications Coordinator, 238 Kent Road, Springfield, PA 19064.

2

The Analytical Balance

The analytical balance used in the introductory laboratory is a precision instrument, capable of detecting the weight of an object of 100 g to within ±0.0001 g (±0.1 mg). This is an uncertainty of only 1 per million. Until the 1950s most of these balances were *two-pan* balances, also referred to as *equal-arm* balances. Then the *single-pan,* or *unequal-arm* (sometimes called *constant-load*), balance essentially replaced the two-pan balance. Today the *electronic* balance (also called the *electromagnetic force* balance) is rapidly replacing the mechanical, or single-pan, balance. It uses an electric current to generate a magnetic force which balances the load placed on the balance pan. The current required to generate the force is directly proportional to the mass of the object on the pan.

We shall first describe the determination of mass and weight and then consider the three types of balances mentioned above.

MASS AND WEIGHT

The equal-arm balance is a lever of the first class; that is, the fulcrum (B in Figure 2.1) lies between the points of application of forces (A and C in Fig. 2.1). Since the arms are equal in length, $1_1 = 1_2$. Pans are suspended from A and C, and the object to be weighed (mass M_1) is placed on the left-hand pan while known weights (mass M_2) are placed on the right-hand pan. Both M_1 and M_2 are attracted by the earth (gravity), the forces being, according to Newton's second law,

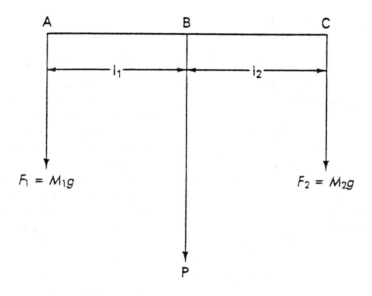

Fig. 2.1 Principle of weighing on a two-pan balance.

$$F_1 = M_1 g$$

$$F_2 = M_2 g$$

where g is the acceleration of gravity. The operator adjusts the value of M_2 until the pointer, P, returns to its original position. Then, according to the principle of moments,

$$F_1 l_1 = F_2 l_2$$

and since $l_1 = l_2$, $F_1 = F_2$, and hence

$$M_1 g = M_2 g \quad \text{or} \quad M_1 = M_2$$

Since M_2 is known, M_1 is determined.

It should be noted that the *weight* of an object is the force exerted on the object by gravitational attraction (F_1 above). The *mass* is the quantity of matter of which the object is composed. The weight of an object is different at different locations on the earth's surface, whereas the mass is *invariant*. It is mass that the analyst determines, but since the term g cancels in the case of the equal-arm balance, the ratio of weights is the same as the ratio of masses. Hence it is customary to use the term *weight* instead of mass, and we commonly speak of the process of determining the mass of an object *weighing*.

TWO-PAN BALANCES

The beam of a two-pan balance contains three prism-form "knife-edges," located at A, B, and C in Fig. 2.1. The Scottish chemist, Joseph Black (1728–1799) is credited with first introducing the use of knife-edges, which are made of agate, a hard, brittle material. It was only after knife-edges were used that analytical weighing measurements could be attained with the two-pan balance.

Weighing on a two-pan balance was always tedious and time-consuming and scientists constantly sought ways to make the operation easier. In addition, there are two sources of error which are inherent in such a balance. First, in order to obtain accurate results, it is necessary for the two balance arms to be equal in length. In practice, it is very difficult to make beams in which the balance arms are of identical length. If one arm is only 1/100,000 longer than the other, an error of 1/100,000 of the weight is introduced. With a load of 100 g, this amounts to an error or 1 mg. Second, for constant scale deflection per unit of weight added (sensitivity), the three knife edges must lie exactly on a straight line. An increasing load tends to bend the beam slightly, bringing the terminal knife edges below the plane of the central knife edge. This results in a smaller scale deflection per unit weight at higher balance loads.

SINGLE-PAN BALANCES

The two sources of error inherent in the two-pan system are eliminated in the single-pan, or unequal-arm, balance. A schematic diagram of a typical mechanical balance of this type is shown in Fig. 2.2. There are two knife edges rather than three, and the balance arms are unequal in length. The balance pan and a full complement of weights are suspended from the short arm, and the longer arm has a constant counterweight (plus a damping device) built into the balance beam. Hence the empty balance is fully loaded. When the object to be weighed is placed on the pan, weights are removed from the shorter arm by turning knobs on the outside of the balance case. A set of dials shows the sum of the weights removed. Thus weighing is done by *substitution,* leaving the same load on the beam at all times. Constant loading of the beam provides constant balance sensitivity, a characteristic not found in two-pan balances.

Once the sum of weights removed is within 0.1 g (but on the low side) of the weight of the object, the beam is fully released and allowed to come to rest. A reticle, consisting of a scale inscribed on a glass plate, is attached to the longer arm of the balance. The deflection of the beam from its original point of rest is greatly magnified by projecting optically the image of this scale onto a glass plate on the front of the balance. The scale displacement in milligrams can be read directly and some device, such as a vernier, is employed to read to the nearest 0.1 mg.

A number of commercial balances contain so-called *tare* devices, which enable the operator to set the balance at zero with an empty weighing bottle, or beaker,

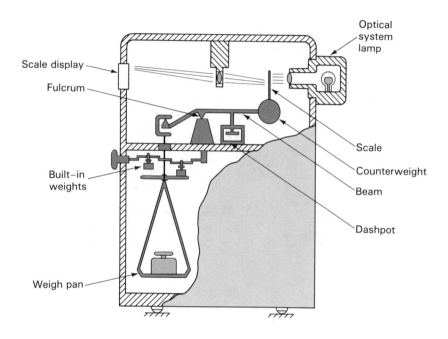

Fig. 2.2 A modern single-pan mechanical analytical balance. [Reprinted from R. M. Schoonover, *Anal. Chem.*, 54, 973A (1982). Reprinted by permission of the author and the American Chemical Society.]

on the pan. The operator can then weigh the sample by pouring it directly into the container and avoid the operation of subtracting the weight of the container from that of the container plus sample.

ANALYTICAL WEIGHTS

A series of reference masses, called *weights*, is employed along with a two-pan balance in determining the mass of an object. These weights are kept in separate compartments in a box, as shown in Fig. 2.3, and are handled with a pair of ivory-tipped forceps. The box is usually lined with velvet to prevent the weights from becoming scratched. The weights above 1 g are usually made of brass or bronze, and are plated with a metal such as chromium or are coated with lacquer. At the present time, increasing use is being made of alloys that are nonmagnetic and resistant to corrosion, such as iron, chromium, and nickel. The large weights normally provided in a set are as follows: 1-, 2-, 2'-, 5-, 10-, 10'-, 20-, 50-, and sometimes a 100-g unit. The weights less than 1 g, called *fractionals*, are usually made of aluminum (sometimes platinum), and are furnished in units of 500, 200,

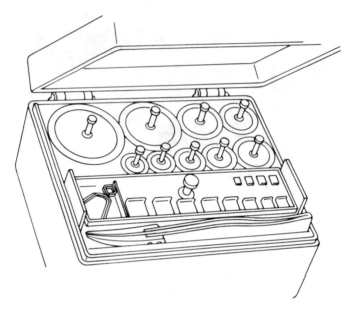

Fig. 2.3 Analytical weights.

TABLE 2.1 TOLERANCES FOR ANALYTICAL WEIGHTS (mg)

Denomination	Class M Individual	Group†	Class S Individual	Group†	Class S-1	Class P
50 g	0.25		0.12		0.60	1.2
30	0.15		0.074 ⎫		0.45	0.90
20	0.10		0.074 ⎬ 0.154		0.35	0.70
10	0.050		0.074 ⎭		0.25	0.50
5	0.034 ⎫		0.054 ⎫		0.18	0.36
3	0.034 ⎬ 0.065		0.054 ⎬ 0.105		0.15	0.30
2	0.034 ⎪		0.054 ⎪		0.13	0.26
1	0.034 ⎭		0.054 ⎭		0.10	0.20
500 mg	0.010 ⎫		0.025 ⎫		0.080	0.16
300	0.010 ⎬ 0.020		0.025 ⎬ 0.055		0.070	0.14
200	0.010 ⎪		0.025 ⎪		0.060	0.12
100	0.010 ⎭		0.025 ⎭		0.050	0.10
50	0.010 ⎫		0.014 ⎫		0.042	0.085
30	0.010 ⎬ 0.020		0.014 ⎬ 0.034		0.038	0.075
20	0.010 ⎪		0.014 ⎪		0.035	0.070
10	0.010 ⎭		0.014 ⎭		0.030	0.060

†No combination of weights in this group shall differ from the sum of the nominal values by more than the amount listed.

Source: Taken from Nat. Bur. Standards Circ. 547 (1954); errata sheet issued April 10, 1962, changed individual tolerances for class M weights below 1 g to 0.010 from 0.054; group tolerances are changed from 0.0105 to 0.020.

100, 100', 50, 20, 10, 10', and sometimes a 5-mg unit. Duplicate values are always marked in some manner such as stamping them with one or two dots. It is possible to obtain sets containing 50-, 30-, 20-, 10-, 5-, 3-, 2-, and 1-g weights. With this distribution no duplication of pieces occurs. Riders are usually made of aluminum, occasionally of platinum. They usually weigh 10 mg, although occasionally riders weighing 5, 6, or 12 mg are encountered.

The National Institute of Standards and Technology classifies analytical weights in several categories. Class M weights are of one-piece construction and are generally reserved for reference standards or for work requiring high precision. They are frequently used to check the calibrations of the built-in weights of a single-pan balance. Class S weights may consist of a base into which the top or handle is screwed, and are used for most calibration work. Class S-1 weights are generally used for routine analytical work and class P (formerly S-2) weights are used for routine laboratory work. A list of acceptable tolerances set by the Bureau of Standards is given in Table 2.1 for these four classes of weights.

ELECTRONIC BALANCES[1]

The *electronic,* or *electromagnetic force,* balance is based on the principle that when a current is passed through a wire placed between the poles of a permanent magnet, a force is generated which moves the wire outside the magnetic air gap. The application of this principle in an analytical balance is best understood by examining an *electromagnetic servo system* shown schematically in Fig. 2.4. In this system the force associated with the object being weighed is coupled mechanically to a servo motor that generates the opposing magnetic force. The system contains a null, or position, indicator which checks the position of the wire in the magnetic field. This may be an optical device, consisting of a vane attached to the beam, a small lamp, and a photo detector. When the two forces are in equilibrium the "error" indicator is at the reference position, and the average current in the servo motor coil is proportional to the resultant force that is holding the mechanism at the reference position. When the beam is displaced from its balanced position, the amount of radiation reaching the detector changes, and causes a very rapid change in current through the coil. An error signal is sent to the circuit that generates a correction current. This current flows through the coil attached to the base of the balance pan, creating a magnetic field and restoring the indicator to its reference position. The correction current needed to restore the system is proportional to the mass of the object on the balance pan. Calibration is performed by placing a known weight on the balance pan and adjusting the circuitry to indicate the mass of the calibrating weight.

Fig. 2.5 shows a schematic diagram of a top-loading electronic balance. No balance beam is needed, but the weight pan is attached to a solid parallelogram

[1]R. M. Schoonover, *Anal. Chem.,* 54, 973A (1982).

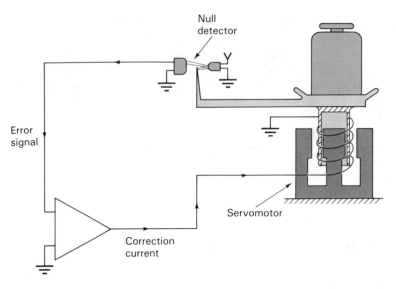

Fig. 2.4 Simplified electromagnetic servo system. [Reprinted from R. M. Schoonover, *Anal. Chem.*, 54, 973A (1982). Reprinted by permission of the author and the American Chemical Society.]

Fig. 2.5 Top-loading balance. [Reprinted from R. M. Schoonover, *Anal. Chem.*, 54, 973A (1982). Reprinted by permission of the author and the American Chemical Scoiety.]

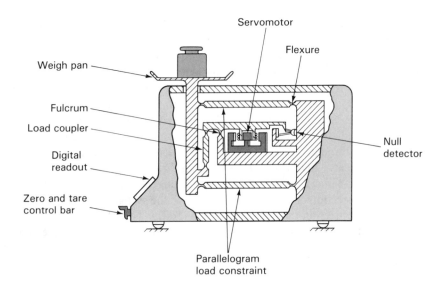

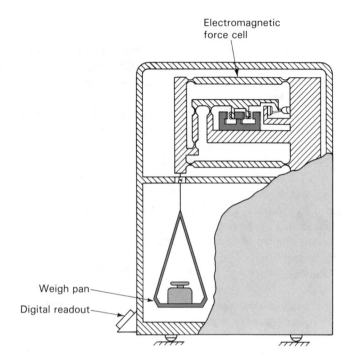

Fig. 2.6 The force balance cell in a classical enclosure. [Reprinted from R. M. Schoonover, *Anal. Chem.*, 54, 973A (1982). Reprinted by permission of the American Chemical Society.]

load constraint to prevent torsional forces, caused by off-center loading, from perturbing the alignment of the balance mechanism. Figure 2.6 shows the force balance cell in a classical balance case, where the weighing pan is placed below the cell rather than above it, as in Fig. 2.5. This configuration gives better axial alignment, or minimum off-center loading, and a reduction in servo motor force with a corresponding drop in capacity.

There are many models of electronic balances available today with numerous optional features. A typical balance will have a way of indicating the zero setting with no load on the pan, and a digital display to give the weight of the object. Some characteristics listed by the Mettler Instrument Corporation for its Model AE100 are as follows: weighing range, 0–109 g; readability, 0.1 mg; reproducibility (s), 0.1 mg; linearity, ±0.2 mg; stabilization time, ~ 5 s; built-in calibration weight, 100 g. The semimicro Model AE240 has a readability of 0.01 mg and a reproducibility of 0.02 mg.

Electronic balances do have limitations and potential errors. These will be discussed in the next section.

ERRORS IN WEIGHING

Students should obviously avoid careless errors in weighing, such as spilling the sample, and misreading the value of a weight, dial, or the position of a vernier. In addition the following possible errors should be avoided.

1. Samples which may take up water vapor or carbon dioxide from the air during the weighing process should be kept in closed containers or weighing bottles. Ignited precipitates are usually weighed in closed crucibles.
2. A glass vessel should not be wiped with a dry cloth before the vessel is weighed. The object may acquire a charge of static electricity and cause an error in weight.
3. The object weighed should be at the same temperature as the balance. Crucibles that have been heated and samples that have been dried should always be cooled to room temperature before weighing.

Electronic Balances

Electronic balances deserve special comments. Schoonover pointed out three possible causes of inaccuracy or imprecision in using such balances: (1) interference by ferromagnetic or magnetized samples; (2) interference by electromagnetic radiation from nearby equipment; and (3) dust which may lodge between the coil and permanent magnet of the servomotor. Johnson and Wells[2] discussed these effects in some detail. They also found that small, but significant, errors may occur if samples are placed off-center on the pan of a top-loading electronic balance.

Finally, it should be noted that physical "weights" are not added or removed when weighing on an electronic balance. This balance measures the force (weight) on the pan directly, and may be more sensitive to buoyancy effects than is a single-pan balance. The effect of buoyancy is discussed in the next section.

Buoyancy

In the normal weighing process, both the object and the weights are buoyed up by the weight of air displaced. This is in accordance with the principle of Archimedes. If the object and weights displaced the same amount of air, no error would be introduced by this effect. This is not usually the case, however, since the density of the weights is normally different from that of the object. In quantitative analysis the density of the weights is usually larger, and hence the object displaces a greater volume of air than do the weights. The weight of the object is therefore less in air (the apparent weight) than it would be in a vacuum (the true weight). The true weight, W_v, is given by

$$W_v = W_a + (V_o - V_w)d_a$$

[2]B. B. Johnson and J. D. Wells, *J. Chem. Ed.,* **63**, 86 (1986).

where W_a is the weight in air, V_o and V_w are the volumes of the object and weights, respectively, and d_a is the density of air (about 0.0012 g/ml under usual conditions).

Since $V_o = W_v/D_o$ or approximately W_a/D_o, and $V_w = W_a/D_w$, where D_o and D_w are the densities of the object and weights, respectively,

$$W_v = W_a + \left(\frac{W_a}{D_o} - \frac{W_a}{D_w}\right)0.0012$$

or

$$W_v = W_a\left[1 + \left(\frac{1}{D_o} - \frac{1}{D_w}\right)0.0012\right]$$

In the usual weighings made in the analytical laboratory, the errors caused by the buoyancy are quite small. Most analytical results are expressed in terms of the ratio of two weights. If the densities of the sample and final precipitate are nearly equal, no appreciable error is introduced by using weights in air. In such operations as the calibration of volumetric apparatus by weighing large volumes of water, the error is appreciable since the density of water is much less than that of the weights.

The following example illustrates the magnitude of the error due to buoyancy in weighing a liquid sample.

Example 1. A sample of benzene, density 0.88 g/ml, is pipetted into an empty glass vessel which weighs 12.2480 g. The bottle plus benzene is found to weigh 14.46204 g. A single-pan balance with stainless steel weights, density 7.8 g/ml, is used. Calculate the true weight (W_v) of the benzene.

The weight in air, W_a, is 14.4624 − 12.2480 = 2.2144 g. Substituting in the expression above,

$$W_v = 2.2144 \left[1 + \left(\frac{1}{0.88} - \frac{1}{7.8}\right)0.0012\right]$$

$$W_v = 2.2144 \left[1 + (1.1364 - 0.1282)\,0.0012\right]$$

$$W_v = 2.2171 \text{ g}$$

Direct reading electronic balances are calibrated by the manufacturer to indicate what is known as "apparent mass vs. 8.0 g/ml." The calibration procedure for such a balance is essentially the setting of its sensitivity so that the apparent mass in air of a standard "weight" of density 8.0 g/ml is indicated correctly. Since the force of gravity at the factory may not be the same as that in the user's laboratory, the user should calibrate his balance with a standard mass in his own laboratory. This can be done on a balance such as the Mettler AE100 using the built-in calibration weight which is automatically lowered by a lever. If the operator so desires, he can use his own external standard mass for calibration.

3

Acid-Base Titrations

The principles of acid-base reactions are discussed in Chapters 3, 6, and 7 of the text. Directions are given in this chapter for several of the laboratory exercises used in introductory courses in quantitative analysis. A brief discussion of some of the reagents is also included.

ACID-BASE REAGENTS

In laboratory practice it is customary to prepare and standardize one solution of an acid and one of a base. These two solutions can then be used to analyze unknown samples of acids or bases. Since acid solutions are more easily preserved than basic solutions, an acid is normally chosen as a permanent reference standard in preference to a base.

In choosing an acid to use in a standard solution, the following factors should be considered. (1) The acid should be strong, that is, highly dissociated. (2) The acid should not be volatile. (3) A solution of the acid should be stable. (4) Salts of the acid should be soluble. (5) The acid should not be a sufficiently strong oxidizing agent to destroy organic compounds used as indicators.

Hydrochloric and sulfuric acids are most widely employed for standard solutions, although neither satisfies all the foregoing requirements. The choride salts of silver, lead, and mercury (I) ion are insoluble as are the sulfates of the alkaline earth metals and lead. This does not normally lead to trouble, however, in most applications of acid-base titrations. Hydrogen chloride is a gas, but is not appreciably volatile from solutions in the concentration range normally employed because it is so highly dissociated in aqueous solution. A solution as concentrated as

0.5 N can be boiled for some time without losing hydrogen chloride if the solution is not allowed to concentrate by evaporation. Nitric acid is seldom used, because it is a strong oxidizing agent, and its solutions decompose when heated or exposed to light. Perchloric is a strong acid, nonvolatile and stable toward reduction in dilute solutions. The potassium and ammonium salts may precipitate from concentrated solutions when formed during a titration. Perchloric acid is commonly preferred for nonaqueous titrations. It is inherently a stronger acid than hydrochloric and is more strongly dissociated in an acidic solvent, such as glacial acetic acid.

Sodium hydroxide is the most commonly used base. Potassium hydroxide offers no advantage over sodium hydroxide and is more expensive. Sodium hydroxide is always contaminated by small amounts of impurities, the most serious of which is sodium carbonate. When CO_2 is absorbed by a solution of NaOH, the following reaction occurs:

$$CO_2 + 2OH^- \longrightarrow CO_3^{2-} + H_2O$$

Carbonate ion is a base but it combines with hydrogen ion in two steps:

$$CO_3^{2-} + H_3O^+ \longrightarrow HCO_3^- + H_2O \tag{1}$$

$$HCO_3^- + H_3O^+ \longrightarrow H_2CO_3 + H_2O \tag{2}$$

If phenolphthalein is employed as the indicator, the color change occurs when reaction (1) is complete; that is, the carbonate ion has reacted with only one H_3O^+ ion. This results in an error, since two OH^- ions were used in the formation of one CO_3^{2-}. If methyl orange is used as the indicator, the color change occurs when reaction (2) is complete and no error occurs, since each CO_3^{2-} ion combines with two H_3O^+ ions. However, in the titration of weak acids, phenolphthalein is the proper indicator to use, and if CO_2 has been absorbed by the titrant an error will occur.

There are several ways to minimize the "carbonate error." Barium hydroxide can be used as the titrant. If CO_2 is absorbed by a solution of this base, a precipitate of barium carbonate is apparent:

$$Ba^{2+} + 2OH^- + CO_2 \longrightarrow BaCO_3(s) + H_2O$$

Since barium hydroxide is of limited solubility in water, solutions cannot be more concentrated than about 0.05 N.

The most common method used to avoid the carbonate error is to prepare carbonate-free sodium hydroxide and then protect the solution from the uptake of CO_2 from the air. Carbonate-free sodium hydroxide can be readily prepared from a concentrated solution of the base, one which is about 50% by weight NaOH. Sodium carbonate is insoluble in the concentrated NaOH solution and settles to the bottom of the container. The solution is decanted from the solid Na_2CO_3 and diluted to the desired concentration. It is then stored in a bottle equipped with a tube containing a solid material (soda lime or Ascarite) which absorbs CO_2 from any air that enters.

The acid-base solutions used in the laboratory are usually in the concentration range of about 0.05 to 0.5 N, most often about 0.1 N. Solutions of such concentrations require reasonable volumes (30 to 50 ml) for titration of samples which are of convenient size to weigh on the analytical balance. For example, 0.6000 g of a pure substance of equivalent weight 200 will require 30 ml of a 0.1 N solution for titration.

Primary Standards

In laboratory practice it is customary to prepare solutions of an acid and a base of approximately the desired concentration and then to standardize the solutions against a primary standard. It is possible to prepare a standard solution of hydrochloric acid by direct weighing of a portion of constant-boiling HCl of known density, followed by dilution in a volumetric flask. More frequently, however, solutions of this acid are standardized in the customary manner against a primary standard.

The reaction between the substance selected as a primary standard and the acid or base should obviously fulfill the requirements for titrimetric analysis. In addition, the primary standard should have the following characteristics:

1. It should be readily available in a pure form or in a state of known purity. In general, the total amount of impurities should not exceed 0.01 to 0.02%, and it should be possible to test for impurities by qualitative tests of known sensitivity.

2. The substance should be easy to dry and should not be so hygroscopic that it takes up water during weighing. It should not lose weight on exposure to air. Salt hydrates are not normally employed as primary standards.

3. It is desirable that the primary standard have a high equivalent weight in order to minimize the consequences of errors in weighing.

4. It is preferable that the acid or base be strong, that is, highly dissociated. However, a weak acid or base may be employed as a primary standard with no great disadvantage, especially when the standard solution is to be used to analyze samples of weak acids or bases.

There are a large number of substances which are suitable for use as primary standards for acid and base solutions. Only a few of the more widely used compounds will be discussed here.

Potassium Acid Phthalate. This compound is the monopotassium salt of phthalic acid ($KHC_8H_4O_4$, abbreviated KHP):

It is readily available in purity of 99.95% or better frm the National Institute of Standards and Technology and from chemical supply houses. It is stable on drying, nonhygroscopic, and has a high equivalent weight, 204.2. It is a weak, monoprotic acid, but since base solutions are frequently used to determine weak acids, this is no disadvantage. Phenolphthalein indicator is employed in the titration, and the base solution should be carbonate-free.

Sulfamic Acid. The formula of this compound is HSO_3NH_2. Sulfamic acid is a strong, monoprotic acid, and either phenolphthalein or methyl red indicator can be employed in the titration with a strong base. It is readily available, inexpensive, and easily purified by recrystallization from water. It is a white crystalline solid, nonhygroscopic, and stable at temperatures up to 130°C. Its equivalent weight is 97.09, considerably smaller than that of KHP. However, the weight which would be normally employed to standardize solutions 0.1 N or greater is sufficiently large to keep weighing errors small. Sulfamic acid is readily soluble in water, and most of its salts are soluble. Solutions of the acid slowly hydrolyze to ammonium acid sulfate. However, the total replaceable hydrogen is unchanged, and this is no disadvantage. Sulfamic acid is generally regarded as the best of the primary standard acids.

Potassium Hydrogen Iodate. This compound, $KH(IO_3)_2$, a strong monoprotic acid, is also an excellent primary standard for base solutions. It is readily available in a form sufficiently pure for use as a primary standard. It is a white, crystalline, nonhygroscopic solid, and it has a high equivalent weight, 389.91. It is sufficiently stable to be dried at 110°C.

Sodium Carbonate. Sodium carbonate, Na_2CO_3, is the most widely used primary standard for solutions of strong acids. It is readily available in a very pure state, except for small amounts of sodium bicarbonate. The bicarbonate can be converted completely into carbonate by heating the substance to constant weight at 270 to 300°C. Sodium carbonate is somewhat hygroscopic but can be weighed without great difficulty. The carbonate can be titrated to sodium bicarbonate, using phenolphthalein indicator, and the equivalent weight is the molecular weight, 106.00. More commonly, it is titrated to carbonic acid using methyl orange indicator. The equivalent weight in this case is one-half the molecular weight, 53.00.

TRIS. The base tris(hydroxymethyl)aminomethane, $(CH_2OH)_3CNH_2$, also called TRIS or THAM, is an excellent primary standard for acid solutions. It is available commercially in purity of 99.95% and is readily dried and weighed. Its reaction with hydrochloric acid is

$$(CH_2OH)_3CNH_2 + H_3O^+ \longrightarrow (CH_2OH)_3CNH_3^+ + H_2O$$

and its equivalent weight is 121.06 g/eq.

Constant-Boiling Hydrochloric Acid. A standard solution of hydrochloric acid can be prepared directly by weighing a portion of the constant-boiling acid and diluting it in a volumetric flask. The constant-boiling acid is prepared by distillation. When the solution is approximately 20% hydrochloric acid by weight, the composition remains unchanged during further distillation. The percentage composition of this solution as a function of the barometric pressure is accurately known. From these data the exact weight of acid in a sample can be calculated and hence the normality obtained. The accuracy that can be attained by this method is quite high. The constant-boiling acid must be prepared with considerable care, and therefore the method is not often used in elementary work. Constant-boiling perchloric acid can also be used to prepare standard solutions.

Gravimetric Standardization. A solution of hydrochloric acid can be standardized by precipitating silver chloride from a known volume of solution and weighing the precipitate. The method assumes, of course, that no chloride ion is present except that furnished by the acid. Highly accurate results can be obtained by this method. Sulfuric acid can be standardized in a similar manner by precipitating and weighing barium sulfate. The barium sulfate precipitate is difficult to obtain in a pure form, however, and the method is not as satisfactory as the gravimetric standardization of hydrochloric acid.

EXPERIMENT 3.1.   **Preparation of 0.1 *N* Solutions of Hydrochloric Acid and Sodium Hydroxide**

Directions are given for the preparation of 1 liter of each solution. If larger volumes or more concentrated solutions are desired, the quantities specified may be increased accordingly.

Procedure

(a) Hydrochloric Acid. Measure into a clean, glass-stoppered bottle approximately 1 liter of distilled water. With a graduated cylinder or measuring pipet, add to the water about 8.5 ml of concentrated hydrochloric acid. Stopper the bottle, mix the solution well by inversion and shaking, and label the bottle.

(b) Sodium Hydroxide. Carbonate-free sodium hydroxide can be prepared most readily from a concentrated solution of the base, because sodium carbonate is insoluble in such a solution. A 1:1 solution of sodium hydroxide in water is available commercially, or may have been prepared by the instructor (Note 1). Carefully add 6 to 7 ml of this solution to approximately 1 liter of distilled water (Note 2) in a clean bottle, using a graduated pipet and rubber bulb. Close the bottle with a rubber stopper (Note 3), shake the solution well, and label the bottle.

Notes

1. If such a solution is not available, dissolve about 50 g of sodium hydroxide in 50 ml of water in a small, rubber-stoppered Erlenmeyer flask. Be careful in handling this solution, as considerable heat is .generated. Allow the solution to stand until the sodium carbonate precipitate has settled. If necessary, the solution can be filtered through a Gooch crucible. Alternatively, carbonate-free base can be prepared by dissolving 4.0 to 4.5 g of sodium hydroxide in about 400 ml of distilled water and adding 10 ml of 0.25 M barium chloride solution. The solution is well mixed and then allowed to stand overnight so that barium carbonate will settle out. The solution is then decanted from the solid into a clean bottle and diluted to 1 liter.

2. Some directions call for boiling the water for about 5 min to remove carbon dioxide. If this is done (consult the instructor), be sure to protect the water from the atmosphere as it cools.

3. Glass-stoppered bottles should not be used since alkaline solutions cause the stoppers to stick so tightly that they are difficult or impossible to remove. Polyethylene bottles, if available, are excellent for storing dilute base solutions. It may be desirable to protect the solution from atmospheric carbon dioxide (consult the instructor). This can be done by fitting a two-holed rubber stopper with a siphon and soda-lime tube, as shown in Fig. 3.1.

EXPERIMENT 3.2. **Determination of the Relative Concentrations of the Hydrochloric Acid and Sodium Hydroxide Solutions**

In this experiment the ratio of the concentrations of the acid and base solutions is determined. Following standardization of either solution, the normality of the other can be calculated from this ratio.

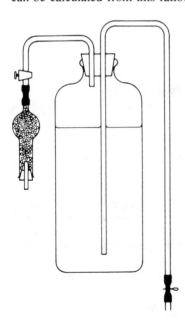

Fig. 3.1 Bottle for storing carbonate-free base.

Procedure

Rinse two clean burets and fill one with the hydrochloric acid and the other with the sodium hydroxide solution prepared in Experiment 3.1. Remove any air bubbles from the tips, lower the liquid level to the graduated portions, and record the initial reading of each buret.

Now run about 35 to 40 ml of the hydrochloric acid solution into a clean 250-ml Erlenmeyer flask and record the buret reading (Note 1). Add 2 drops of phenolphthalein indicator (Note 2) and about 50 ml of water from a graduated cylinder, rinsing down the walls of the flask. Now run into the flask the sodium hydroxide solution from the other buret, swirling the flask gently and steadily to mix the solutions. As an aid in preventing overrunning of the end point, notice the transient, local, pink coloration as it becomes more persistent with the progress of the titration. Finally, when the color first pervades the entire solution even after thorough mixing, stop the titration and record the buret reading. The color should persist for at least 15 s or so, but may gradually fade because of the absorption of atmospheric carbon dioxide. It is well to rinse down the inside of the flask and also the buret tip with distilled water just before the termination of the titration so that stray droplets will not escape the reaction. If the end point is accidentally overrun, the titration can still be salvaged: run enough hydrochloric acid solution into the flask to turn the phenolphthalein indicator colorless, record the buret reading again, and then approach the end point once more with the sodium hydroxide solution.

Repeat the titration at least two more times (Note 3). Finally, calculate the volume of acid equivalent to 1 ml of base:

$$1.000 \text{ ml of base} = \frac{\text{volume of acid}}{\text{volume of base}}$$

Use buret corrections if necessary (consult the instructor).

Notes

1. This volume is recommended in order to minimize errors in reading a buret. The acid is usually titrated with base instead of base with acid to minimize absorption of carbon dioxide during the titration.

2. Methyl red, bromthymol blue, and other indicators can also be employed. Solutions are prepared as follows: *Phenolphthalein:* Dissolve 2 g of the phenolphthalein per liter of 95% ethanol. *Methyl red:* Dissolve 1 g of the sodium salt of methyl red per liter of water. *Bromthymol blue:* Dissolve 0.1 g of bromthymol blue in 8 ml of 0.02 M NaOH and dilute to 100 ml with water.

3. Do not allow the sodium hydroxide to remain in the buret any longer than nec-

essary. As soon as the titrations are finished, drain the base from the buret and rinse thoroughly, first with dilute hydrochloric acid and then with water.

EXPERIMENT 3.3. Standardization of Sodium Hydroxide Solution with Potassium Acid Phthalate

As mentioned above, a number of good primary standards are available for standardizing base solutions. Directions are given here for the use of potassium acid phthalate, but these can be readily modified to suit another standard.

Procedure

Place about 4 to 5 g of pure potassium acid phthalate in a clean weighing bottle and dry the sample in an oven at 110°C for at least 1 h. Cool the bottle and its contents in a desiccator (Note 1). Weigh accurately into each of three clean, numbered Erlenmeyer flasks about 0.7 to 0.9 g of the potassium acid phthalate (Note 2). Record the weights in your notebook.

To each flask add 50 ml of distilled water (Note 3) from a graduated cylinder, and shake the flask gently until the sample is dissolved. Add 2 drops of phenolphthalein to each flask. Rinse and fill a buret with the sodium hydroxide solution. Titrate the solution in the first flask with sodium hydroxide to the first permanent pink color. Your hydrochloric acid solution, in a second buret, may be used for back-titration if required. Repeat the titration with the other two samples, recording all data in your notebook.

Calculate the normality of the sodium hydroxide solution obtained in each of the three determinations. Average these values and compute the average deviation in the usual manner. If the average deviation exceeds about 2 parts per thousand, consult the instructor. Finally, calculate the normality of the hydrochloric acid solution from the normality of the sodium hydroxide and the volume ratio of the acid and base obtained in Experiment 3.2.

Notes

1. Potassium acid phthalate is relatively nonhygroscopic, and the drying process may be omitted (consult the instructor).
2. Since 4 meq of potassium acid phthalate weighs $4 \times 204.2 = 816.8$ mg, the quantity recommended should require 35 to 45 ml of a $0.1\ N$ base solution for titration.
3. Some directions call for boiling the water for about 5 min to remove carbon dioxide before use. If this is done (consult the instructor), the water should be cooled to room temperature before the titration, and it should be protected from the atmosphere while cooling.

EXPERIMENT 3.4. **Standardization of the Hydrochloric Acid Solution with Sodium Carbonate**

The hydrochloric acid solution can be standardized against a primary standard if so desired. Sodium carbonate is a good standard and is particularly recommended if the acid solution is to be used to titrate carbonate samples.

Procedure

Accurately weigh three samples (about 0.20 to 0.25 g each) of pure sodium carbonate (Note 1), which has been previously dried, into three Erlenmeyer flasks. Dissolve each sample with about 50 ml of distilled water and add 2 drops of methyl red[1] or methyl orange (see Note 2 and consult the instructor).

(a) Methyl Red. Titrate each sample with the hydrochloric acid solution. Methyl red is yellow in basic and red in acid solution. As soon as the solution is distinctly red, add 1 additional ml of hydrochloric acid and remove the carbon dioxide by boiling the solution gently for about 5 min (Note 3). Cool the solution to room temperature and complete the titration. Back-titration can be done with the sodium hydroxide solution previously prepared. If the color change is not sharp, repeat the heating to remove carbon dioxide.

(b) Methyl Orange. Prepare a solution of *p*H 4 by dissolving 1 g of potassium acid phthalate in 100 ml of water. Add 2 drops of methyl orange to this solution and retain it for comparison purposes. Now titrate each sample with hydrochloric acid until the color matches that of the comparison solution.

Calculate the normality of the hydrochloric acid solution obtained in each of the three titrations. Average these values and compute the average deviation in the usual manner. If this figure exceeds 2 to 3 parts per thousand, consult the instructor. The normality of the sodium hydroxide solution can be calculated from the normality of the acid and the relative concentrations of acid and base obtained in Experiment 3.2.

Notes

1. Analytical-grade sodium carbonate (assay value 99.95%) can be used after drying for about $\frac{1}{2}$ h at 270 to 300°C.

[1] The color changes shown by these indicators can be modified by addition of suitable dye. Methyl red is frequently mixed with methylene blue, the color change then being red-violet (acid) to green (base) with an intermediate shade of gray. The dye xylene cyanole FF is added to methyl orange to give "modified methyl orange." The color change is then pink (acid) to green (base) with a gray intermediate. See I. M. Kolthoff and C. B. Rosenblum, *Acid-Base Indicators*, Macmillan Publishing Co., Inc., New York, 1937.

2. Various indicators and mixed indicators have been suggested for this titration. The *p*H at the equivalence point of the reaction

$$CO_3^{2-} + 2H^+ \rightleftharpoons H_2CO_3$$

is about 4, and methyl orange changes color near this *p*H. The titration curve is not very steep, however, and hence it is often suggested that excess acid be added and carbon dioxide removed by boiling or vigorous shaking. The subsequent titration of excess acid with base involves only strong electrolytes, and a sharp end point is obtained if the carbon dioxide is completely removed. An indicator blank must then be determined since methyl orange changes color at a *p*H appreciably different from 7. If methyl red (*p*H 5.4) is employed, no indicator blank is necessary. Directions are given here for the titration to the methyl red end point with removal of carbon dioxide and for the titration to the methyl orange end point without removal of carbon dioxide. The latter procedure is more rapid and is recommended where a high degree of accuracy is not required.

3. If insufficient acid is present to completely convert bicarbonate into carbonic acid, the indicator will turn back to its basic color as the carbon dioxide is expelled and the *p*H rises. The titration is then continued with acid. If excess acid is added, the indicator will retain the acid color and the titration is continued by addition of base.

EXPERIMENT 3.5.   Determination of the Purity of Potassium Acid Phthalate

Directions are given for the titration of a solid acid sample. Commercial unknowns usually contain potassium acid phthalate or sulfamic acid, either of which can be titrated with phenolphthalein indicator.

Procedure

Dry the unknown sample in a weighing bottle (unless otherwise directed) at 110°C for at least 1 h, and cool in a desiccator. Weigh three samples of appropriate size (Note 1) into clean Erlenmeyer flasks and dissolve in 75 to 100 ml of distilled water (Note 2). Add 2 drops of phenolphthalein indicator to each flask.

Titrate the contents of the first flask with the standard sodium hydroxide solution to the first permanent pink color. Record all buret readings as usual and then titrate the other two samples in the same manner.

Calculate the percentage of potassium acid phthalate in each sample and obtain the average purity and average deviation (Note 3). Report the percentage purity as suggested on page 30.

Notes

1. The instructor will specify the size sample required to use 30 to 45 ml of a 0.1 *N* base for titration. If this weight is greater than 1 g, as it may be with potassium acid phthalate samples of low percentage purity, the samples need be weighed only to the nearest milligram. Why?

2. Some directions call for boiling the water for about 5 min to remove carbon dioxide

before use. If this is done (consult instructor), the water should be cooled to room temperature before the titration, and it should be protected from the atmosphere while cooling.

3. The precision that can be obtained depends upon the homogeneity of the sample as well as the technique of the student. Consult the instructor for the expected precision.

EXPERIMENT 3.6.   Determination of Acetic Acid Content of Vinegar

The principal acid in vinegar is acetic acid, and federal standards require at least 4 g of acetic acid per 100 ml of vinegar. The total quantity of acid can be readily determined by titration with standard base using phenolphthalein indicator. Although other acids are present, the result is calculated as acetic acid.

Procedure

Pipet 25 ml of vinegar into a 250-ml volumetric flask, dilute to the mark, and mix thoroughly (Note). Pipet a 50-ml aliquot of this solution into an Erlenmeyer flask and add 50 ml of water and 2 drops of phenolphthalein indicator. Titrate with standard base to the first permanent pink color. Repeat the titration on two additional aliquots.

Assuming all the acid to be acetic, calculate the number of grams of acid per 100 ml of vinegar solution. Assuming that the density of vinegar is 1.000, what is the percentage of acetic acid by weight in vinegar? Average your results in the usual manner.

Note

This quantity should require a reasonable volume of 0.1 N base for titration. If a more concentrated base solution is employed, a 100-ml volumetric flask can be used. The dilution with water prevents the color of the vinegar solution from interfering in the detection of the end point.

EXPERIMENT 3.7.   Determination of the Alkalinity of Soda Ash

Crude sodium carbonate, called soda ash, is commonly used as a commercial neutralizing agent. The titration with standard acid to the methyl orange end point gives the total alkalinity, which is mainly due to sodium carbonate. Small amounts of sodium hydroxide and sodium bicarbonate may also be present. The results are usually expressed as a percentage of sodium carbonate or sodium oxide.

Since the samples are frequently nonhomogeneous, the method of aliquot portions is employed. Either methyl red or methyl orange can be employed as the indicator.

Procedure

Weigh accurately into a clean 250-ml beaker a sample of the dried unknown of appropriate size (Note). Dissolve the sample in about 125 ml of distilled water. Place a clean funnel in a 250-ml volumetric flask and transfer the solution from the beaker to the flask. Rinse the beaker, add the rinsing to the flask, and finally dilute to the mark. Mix the contents of the flask thoroughly by inversion and shaking.

Pipet a 50-ml aliquot into an Erlenmeyer flask and add 2 drops of methyl red or methyl orange. Titrate with standard acid according to procedure (a) or (b), page 50. Repeat the titration with two other 50-ml aliquots. At the end of the titrations, be sure to empty and thoroughly rinse the volumetric flask. An alkaline solution should not be left in a volumetric flask for a long period of time.

Report the percentage of sodium carbonate or sodium oxide (consult the instructor) in the sample. A precision of 3 to 5 parts per thousand is not unusual for the titration.

Note

The instructor will specify the size of sample required to use 30 to 45 ml of 0.1 N acid for titration.

EXPERIMENT 3.8.   Titration of Antacids

The purpose of this experiment is to determine the amount of acid neutralized by various commercial antacids. These products contain bases such as calcium carbonate, magnesium carbonate, and magnesium hydroxide. The latter compounds are not very soluble in water, but a direct titration can be carried out with hydrochloric acid if sufficient time is allowed for reaction between the solid and the titrant. A recurring end point may be obtained because this reaction is rather slow. The pH of a solution made by dissolving the antacid in water is in the range 8 to 9, indicating the presence of bicarbonate ion. Hence methyl orange should be used as the indicator in a direct titration.

In the procedure below, excess acid is added to react with the antacid, the solution is heated to remove CO_2, and the excess acid is titrated with standard base. Phenolphthalein can be used as the indicator and a reasonably sharp end point is obtained.

Procedure

Cut a tablet of the antacid in half and weigh one piece on the analytical balance (Note 1). Transfer the sample to a 250-ml Erlenmeyer flask; add 75 ml of standard 0.1 *N* HCl solution using a pipet. Heat the solution to boiling and boil it gently for about 3 min (Note 2). Cool the solution to room temperature, add 4 drops of phenolphthalein indicator, and titrate with standard base to the first permanent pink color.

(a) Calculate the grams of HCl neutralized by 1 g of the antacid. (b) Calculate the grams of 0.1 *N* hydrochloric acid solution neutralized by 1 g of the antacid. (Assume that 0.1 *N* HCl has a density of 1.00 g/ml.) You may have heard advertising claims about how many times its weight of "stomach acid" an antacid neutralizes. Which weight is referred to in such claims, the one calculated in (a) or in (b) above?

Notes

1. Tablets of such products as Tums and Rolaids weigh about 1.3 to 1.4 g and require 80 to 110 ml of 0.1 *N* HCl for direct titration. If 0.2 *N* standard acid and base are at hand the entire tablet can be titrated.

2. There will probably be a small amount of white solid (filler) which does not dissolve even after heating.

EXPERIMENT 3.9.   Titrations in Nonaqueous Media

The principles of titrations in media other than water are discussed in Chapter 6 of the text. Such titrations are rather widely used today, and it is appropriate to include one or two in a beginning course in volumetric analysis. The procedures given below employ a solution of perchloric acid in glacial acetic acid. This solution can be standardized against pure potassium acid phthalate which acts as a base in acetic acid solvent. The solution is then used to titrate an amine which is too weak a base to be titrated in water. Methyl violet is used as the indicator, or the end point can be detected potentiometrically using a glass and calomel electrode pair. These directions are based upon the recommendations of Fritz.[2]

Procedure

Preparation of Solutions

(a) 0.1 *N HClO₄*. Add 4.3 ml of 72% perchloric acid to 150 ml of glacial acetic acid, mix well, add 10 ml of acetic anhydride, and allow the solution to stand for 30 min (Note 1). Dilute to 500 ml with glacial acetic acid and allow the solution to cool to room temperature.

(b) 0.1 *N sodium acetate*. Dissolve 4.1 g of anhydrous sodium acetate in glacial acetic acid and dilute to 500 ml with the acid.

(c) *Methyl violet indicator*. This is prepared by dissolving 0.2 g of methyl violet in 100 ml of chlorobenzene.

---

[2]J. S. Fritz, *Acid-Base Titrations in Nonaqueous Solutions*, G. F. Smith Chemical Co., Columbus, Ohio, 1952.

Relative Concentrations. Fill two burets with the perchloric acid and sodium acetate solutions (Note 2). Withdraw about 35 to 40 ml of perchloric acid, add 2 drops of methyl violet indicator, and titrate with the sodium acetate solution. Take the first permanent violet tinge as the end point. Repeat the titration with two additional samples and calculate the relative concentrations of the two solutions.

Standardization. Weigh samples of about 0.5 to 0.6 g of pure potassium acid phthalate into three Erlenmeyer flasks and add about 60 ml of glacial acetic acid to each flask. Heat the first flask cautiously until the sample is in solution. Then cool and add 2 drops of methyl violet. Titrate with perchloric acid solution to the disappearance of the violet tinge. Use the sodium acetate solution for back-titration if needed.

Repeat the titration with the other two samples and then calculate the normality of the perchloric acid solution. From the relative concentration obtained above, calculate the normality of the sodium acetate solution (Note 3).

Analysis. Amino acids or amines can be titrated with perchloric acid in acetic acid. If the sample is an amino acid, it is recommended that excess perchloric acid be added and the excess titrated with sodium acetate. Amines can be titrated directly with perchloric acid.

Weigh three samples of the substance to be determined, taking about 3 meq for each sample (consult the instructor). If the sample is an amino acid, dissolve in exactly 50 ml (pipet) of standard perchloric acid. Add 2 drops of methyl violet indicator and back-titrate with sodium acetate, taking the first permanent violet tinge as the end point. If the sample is an amine, dissolve it in about 50 ml of acetic acid and titrate with standard perchloric acid to the first appearance of the violet color.

Calculate the number of milliequivalents of amino acid or amine in the sample, and report as desired by the instructor (Note 4).

Notes

1. The acetic anhydride is added to react with water in the perchloric acid.

2. Avoid contact of acetic acid solutions with the skin. If these solutions are spilled on the hands, rinse the hands immediately with tap water.

3. Room temperature should be noted when these titrations are carried out. If there is a large change in temperature between the time of standardization and analysis, a correction should be applied to the volume of acetic acid solution.

4. The instructor may prefer to give pure samples of the amine or amino acid as unknowns. The student then reports the equivalent weight of the unknown.

4

Oxidation-Reduction Titrations

The principles of oxidation-reduction reactions are discussed in Chapters 10 and 11 of the text. In the introductory laboratory the oxidizing agents used most commonly in standard solutions are potassium permanganate, potassium dichromate, cerium(IV) sulfate, and iodine. The chemical properties of each of these reagents are discussed in some detail in this chapter and directions for a number of laboratory exercises are given. Other oxidizing agents which may be encountered include periodic acid (H_5IO_6), potassium iodate (KIO_3), potassium bromate ($KBrO_3$), and calcium hypochlorite ($Ca(OCl)_2$).

Standard solutions of reducing agents are not as widely used in the laboratory as those of oxidizing agents, because most reducing agents are slowly oxidized by oxygen of the air. Sodium thiosulfate is the only common reducing agent that can be kept for long periods of time without undergoing air oxidation. This reagent is used exclusively for iodine titrations, and its properties are discussed along with those of iodine. Other reducing agents that are sometimes employed in standard solutions are listed next.

Iron(II) Ion. Solutions of iron(II) ions, in 0.5 to 1 N sulfuric acid, are only slowly oxidized by air. The normality of the solution should be checked at least daily.

Chromium(II) Salts. The chromium(II) ion is a powerful reducing agent, the standard potential of the reaction

$$Cr^{3+} + e \rightleftharpoons Cr^{2+}$$

being -0.41 V. Solutions are oxidized rapidly by air and extreme care must be employed in their use.

Titanium(III) Salts. These salts are also strong reducing agents, the standard potential of the reaction

$$TiO^{2+} + 2H^+ + e \rightleftharpoons Ti^{3+} + H_2O$$

being $+0.04$ V. Solutions of these salts are readily oxidized by air, but are easier to handle than solutions of chromium(II) salts.

Oxalates and Arsenites. The reactions of sodium oxalate (or oxalic acid) and those of arsenious acid will be discussed along with those of potassium permanganate. Standard solutions of oxalic acid are fairly stable; those of sodium oxalate are much less stable. Neutral or weakly acid solutions of arsenious acid are fairly stable but alkaline solutions are slowly oxidized by air.

POTASSIUM PERMANGANATE

Properties

Potassium permanganate has been widely used as an oxidizing agent for over 100 years. It is a reagent that is readily available, inexpensive, and requires no indicator unless very dilute solutions are used. One drop of 0.1 N permanganate imparts a perceptible pink color to the volume of solution usually used in a titration. This color is used to indicate excess of the reagent. Permanganate undergoes a variety of chemical reactions, since manganese can exist in oxidation states of $+2$, $+3$, $+4$, $+6$, and $+7$. These reactions are summarized below:

$$MnO_4^- + 8H^+ + 5e \rightleftharpoons Mn^{2+} + 4H_2O \quad (E° = +1.51 \text{ V}) \quad (1)$$

This is the reaction that takes place in strongly acid solution (0.1 N or greater). The equivalent weight of potassium permanganate is one-fifth the molecular weight, or 31.61.

$$MnO_4^- + 4H^+ + 3e \rightleftharpoons MnO_2 + 2H_2O \quad (E° = +1.70 \text{ V}) \quad (2)$$

This reaction takes place in less-acid solution, as is shown in the equation. It is predominant in the pH range of about 2 to 12. Here the equivalent weight of permanganate is one-third its molecular weight, or 52.68.

$$MnO_4^- + 8H^+ + 4e \rightleftharpoons Mn^{3+} + 4H_2O \quad (E° = +1.50 \text{ V}) \quad (3)$$

Although the trivalent state of manganese normally is not stable, complexing anions such as pyrophosphate or fluoride will stabilize the ion, forming complexes such as $Mn(H_2P_2O_7)_3^{3-}$. The equivalent weight of permanganate is one-fourth the molecular weight, or 39.51.

$$MnO_4^- + e \rightleftharpoons MnO_4^{2-} \quad (E° = +0.54 \text{ V}) \quad (4)$$

This reaction takes place only in strongly alkaline solution (1 M or so in hydroxide ions). In less basic solution reaction (2) will occur. Barium chloride is normally added to precipitate barium manganate, thus removing the green color of the manganate ion and also preventing further reduction from occurring.

The most common reaction encountered in beginning quantitative analysis is the first, that in strongly acid solution. Permanganate reacts rapidly with many reducing agents according to equation (1), but some substances require heating or addition of a catalyst in order to speed up the reaction. Were it not for the fact that many reactions of permanganate are slow, more difficulties would be encountered in the use of this reagent. For example, permanganate is a strong enough oxidizing agent to oxidize manganese(II) ion to manganese dioxide, according to the equation

$$3Mn^{2+} + 2MnO_4^- + 2H_2O \longrightarrow 5MnO_2 + 4H^+$$

The slight excess of permanganate normally present at the end point of a titration is sufficient to bring about precipitation of some manganese dioxide. Theoretically, the acidity could not be increased sufficiently to prevent this precipitation from occurring. However, since the reaction is slow, manganese dioxide is not normally precipitated at the end point of permanganate titrations in acid solution. This reaction can actually be utilized for the volumetric determination of manganese, however, and is known as the Volhard method.

Special precautions must be taken in preparing permanganate solutions. Manganese dioxide catalyzes the decomposition of permanganate solutions. Traces of manganese dioxide initially present in the permanganate, or formed by the reaction of permanganate with traces of reducing agents in the water, lead to decomposition. Directions normally call for dissolving the crystals, heating to destroy reducible substances, and filtering through asbestos or sintered glass (nonreducing filters) to remove manganese dioxide. The solution must then be standardized, and if kept in the dark and not acidified, its concentration will not change appreciably over a period of several months. Sometimes directions call for adding sodium hydroxide to the solution to prevent decomposition.

Acid solutions of permanganate are not stable, because permanganic acid is decomposed according to the following reaction:

$$4MnO_4^- + 4H^+ \longrightarrow 4MnO_2 + 3O_2 + 2H_2O$$

This is a slow reaction in dilute solutions at room temperature. However, one should never add excess permanganate to a reducing agent and then raise the temperature to hasten oxidation because the foregoing reaction will occur at an appreciable rate.

Standardization

Arsenic(III) Oxide. This compound, As_2O_3, is an excellent primary standard for permanganate solutions. It is stable, nonhygroscopic, and is readily available in a high degree of purity. The oxide is dissolved in sodium hydroxide,

$$As_2O_3 + 6NaOH \longrightarrow 2Na_3AsO_3 + 3H_2O$$

the solution acidified with hydrochloric acid (about 0.5 N),

$$Na_3AsO_3 + 3HCl \longrightarrow H_3AsO_3 + 3NaCl$$

and titrated with permanganate. The reaction at room temperature is slow unless a catalyst is added. Iodine, in various oxidation states, is a suitable catalyst. Potassium iodide, potassium iodate,[1] and iodine monochloride[2] have been investigated as catalysts.

Swift[3] suggested the following mechanism for the catalysis by iodine monochloride in hydrochloric acid solutions:

$$2ICl + H_3AsO_3 + H_2O \longrightarrow I_2 + H_3AsO_4 + 2H^+ + 2Cl^-$$
$$5I_2 + 2MnO_4^- + 10Cl^- + 16H^+ \longrightarrow 10ICl + 2Mn^{2+} + 8H_2O$$

These two reactions have been found to occur rapidly at room temperature. Multiplying the first equation by 5 and adding to the second gives

$$2MnO_4^- + 5H_3AsO_3 + 6H^+ \longrightarrow 2Mn^{2+} + 5H_3AsO_4 + 3H_2O$$

This is, of course, the overall reaction between permanganate and arsenious acid.

Sodium Oxalate. This compound, $Na_2C_2O_4$, is also a good primary standard for permanganate in acid solution. It can be obtained in high degree of purity, is stable on drying, and is nonhygroscopic. The reaction with permanganate is somewhat complex, and even though many investigations have been made, the exact mechanism is not clear. The reaction is slow at room temperature and hence the solution is normally heated to about 60°C. Even at elevated temperature the reaction starts slowly, but the rate increases as manganese(II) ion is formed. Manganese(II) ion acts as a catalyst and the reaction is termed "autocatalytic" since the catalyst is produced in the reaction itself. Manganese(II) ion may exert its catalytic effect by rapidly reacting with permanganate to form manganese of intermediate oxidation states ($+3$ or $+4$), which then rapidly oxidize oxalate ion, returning to the divalent state.

The stoichiometric relation follows the equation

$$5C_2O_4^{2-} + 2MnO_4^- + 16H^+ \longrightarrow 2Mn^{2+} + 10CO_2 + 8H_2O$$

For a number of years analysts employed the procedure recommended by McBride,[4] which called for carrying out the entire titration of oxalate with permanganate at

[1] H. A. Bright, *Ind. Eng. Chem., Anal. Ed.*, **9**, 577 (1937); I. M. Kolthoff, H. A. Laitinen, and J. J. Lingane, *J. Am. Chem. Soc.*, **59**, 429 (1937).

[2] D. E. Metzler, R. J. Myers, and E. H. Swift, *Ind. Eng. Chem., Anal. Ed.*, **16**, 625 (1944).

[3] E. H. Swift, *Introductory Quantitative Analysis*, Prentice Hall, Inc., Englewood Cliffs, N.J., 1950, p. 132.

[4] R. S. McBride, *J. Am. Chem. Soc.*, **34**, 393 (1912).

elevated temperature, titrating slowly, and stirring vigorously. Later, Fowler and Bright[5] made a very thorough investigation of possible errors in the titration. They found some evidence for peroxide formation

$$O_2 + H_2C_2O_4 \longrightarrow H_2O_2 + 2CO_2$$

and found that if the peroxide decomposes before reacting with permanganate, too little permanganate is used and the normality found is high. Fowler and Bright recommended that almost all the permanganate be added rapidly to the acidified (about 1.8 N) solution at room temperature. After the reaction is complete the solution is heated to 60°C and the titration is completed at this temperature. This procedure eliminates the foregoing error. Other possible errors in the procedure were found to be negligible.

Iron Wire. Iron wire of high degree of purity is available as a primary standard. It is dissolved in dilute hydrochloric acid, and any iron(III) ion present is reduced to the iron(II) state. If the solution is then titrated with permanganate, an appreciable amount of chloride ion is oxidized in addition to iron(II) ion. The oxidation of chloride ion by permanganate is slow at room temperature. However, in the presence of iron, the oxidation occurs more rapidly. Although iron(II) ion is the stronger reducing agent and should be oxidized first, chloride ion is oxidized simultaneously with iron(II) ion. It is frequently said that iron "induces" this reaction. No such difficulty is encountered in the oxidation of arsenic(III) oxide or sodium oxalate in hydrochloric acid solution.

A solution of manganese(II) sulfate, sulfuric acid, and phosphoric acid, called "preventive" or Zimmermann-Reinhardt solution, is added to the hydrochloric acid solution of iron before titration with permanganate. Phosphoric acid lowers the concentration of iron(III) ion by formation of a complex which helps force the reaction to completion and also removes the yellow color which iron(III) ion shows in chloride media. The phosphate complex is colorless and the end point is made somewhat clearer.

EXPERIMENT 4.1.   Preparation of a 0.1 N Potassium Permanganate Solution

The permanganate titrations described here will be carried out in acid solution. Hence the equivalent weight of potassium permanganate is one-fifth the molecular weight, or 31.61. Directions are given for the preparation of 1 liter of a 0.1 N solution, thereby requiring 31.61 $\times$ 0.1 or about 3.2 g of salt.

Procedure

Since potassium permanganate solutions are susceptible to decomposition (page 58), special precautions are recommended for preparing the solution if it is to be used over a period of several weeks. If the solution is to be prepared,

[5]R. M. Fowler and H. A. Bright, *J. Res. Nat. Bur. Standards*, **15**, 493 (1935).

standardized, and used the same day, the special precautions are not necessary (Note).

Weigh approximately 3.2 g of a good grade of potassium permanganate and place it in a clean 250-ml beaker. Dissolve the salt by adding 50 ml of water and stirring. Decant the solution into a large beaker and add 50 ml of additional water to dissolve any crystals remaining in the first beaker. Repeat this procedure until all the crystals are dissolved. Dilute the solution to about 1 liter, transfer to a glass-stoppered bottle, and label properly.

If the instructor recommends removal of manganese dioxide, proceed as follows: Before transferring the solution to the bottle, heat it just to boiling and keep it slightly below the boiling point for 1 h. Then allow the solution to cool, and filter it through a sintered glass crucible using suction. Transfer the solution to a glass-stoppered bottle and label properly.

Note

Consult the instructor. The instructor may have prepared in advance a large quantity of stock solution which has stood for a week or so. If the manganese dioxide has settled, the clear solution can be withdrawn through an all-glass siphon, avoiding filtration.

EXPERIMENT 4.2. **Standardization of Potassium Permanganate Solution**

Procedure

(a) Sodium Oxalate. Fowler-Bright method. Weight accurately three samples of about 0.25 to 0.30 g each (Note 1) of the dried salt into clean 500-ml Erlenmeyer flasks. Add 250 ml of dilute sulfuric acid (12.5 ml of concentrated acid diluted to 250 ml) which has previously been boiled 10 to 15 min and then cooled to 24 to 30°C. Swirl the flask until the solid dissolves and then titrate with permanganate. Steadily add the permanganate directly into the oxalate solution (not down the walls of the flask) and stir slowly. Add sufficient permanganate (35 to 40 ml) to come within a few milliliters of the equivalence point, adding this at a rate of about 25 to 35 ml/min. Let the solution stand until the pink color disappears (Note 2) and then heat the solution to 55 to 60°C. Complete the titration at this temperature, adding permanganate slowly until 1 drop imparts to the solution a faint pink color that persists at least 30 s. The last milliliter should be added slowly, allowing each drop to be decolorized before adding another. The solution should be as warm as 55°C at the end of the titration.

To about 300 ml of previously boiled dilute sulfuric acid, add permanganate solution dropwise until the color matches that of the titrated solution. This volume (usually about 0.03 to 0.05 ml) is subtracted from the volume used in the titration.

After titration of three samples, calculate the normality obtained in each titration and average the results. With care, the average deviation should be as small as 2 parts per thousand (consult the instructor).

Notes

1. The equivalent weight of sodium oxalate is 67.00. This weight is sufficient to react with about 35 to 45 ml of a 0.1 *N* solution.
2. This may take 30 to 45 s since the reaction is not instantaneous. If the color does not disappear, indicating excess permanganate, discard the solution and add less permanganate to the next sample.

McBride method. In place of the more lengthy Fowler-Bright procedure, the McBride procedure may be preferred for beginning students.

Weigh accurately three samples of about 0.25 to 0.30 g each of dried sodium oxalate into clean 250-ml Erlenmeyer flasks. Dissolve each sample in about 75 ml of 1.5 *N* sulfuric acid (20 ml of concentrated sulfuric acid added to 400 ml of water). Then heat the first solution almost to boiling (80 to 90°C) and titrate slowly with permanganate with constant swirling. The end point is marked by the appearance of a faint pink color that persists at least 30 s. The temperature should not drop below 60°C during the titration. Titrate the other two solutions in the same manner.

To about 100 ml of the 1.5 *N* sulfuric acid, add permanganate solution dropwise until the color matches that of the titrated solution. This volume should be subtracted from the volume used in the titration.

Calculate the normality of the permanganate solution. The average deviation should be as small as about 2 parts per thousand (consult the instructor).

(b) Arsenic(III) Oxide. Weigh accurately three portions of about 0.2 g each of pure arsenic(III) oxide, previously dried in the oven, into each of three 250-ml Erlenmeyer flasks. Add to each flask 10 ml of a cool sodium hydroxide solution made by dissolving 20 g of sodium hydroxide in 80 ml of water. Allow the flask to stand for 8 to 10 min, stirring the solution occasionally until the sample has completely dissolved. Then add 100 ml of water, 10 ml of concentrated hydrochloric acid, and 1 drop of 0.0025 *M* potassium iodide solution as a catalyst (Note). Titrate with the permanganate solution until a single drop imparts to the solution a faint pink color which persists for at least 30 s. after the liquid is swirled. The last milliliter should be added slowly, allowing each drop to be decolorized before adding another. Finally, run a blank to determine the volume of permanganate required to color the solution and to react with any reducible material in the reagents. This should amount to no more than 1 drop of permanganate.

Titrate the other two samples and calculate the normality of the permanganate solution. The average deviation should be about 1 to 2 parts per thousand.

Note

Potassium iodate or iodine monochloride can be used as catalysts if desired. The potassium iodide solution is prepared by dissolving about 0.4 g in 1 liter of water.

(c) Iron Wire. Pure iron wire, free of rust, can be used as a primary standard for permanganate. Weigh accurately three samples of wire of about 0.2 g each into 150-ml beakers. Add 20 ml of 1 : 1 hydrochloric acid to the first sample and warm the solution on a steam bath or over a low flame until all the iron dissolves.

Since the sample is in hydrochloric acid solution, it is convenient to employ tin(II) chloride as the reducing agent and then add preventive solution before titration with permanganate. Directions for this procedure are given under Experiment 4.5.

Repeat the procedure on the other two samples and calculate the normality of the permanganate solution. The average deviation should be about 2 parts per thousand, or less.

Determinations with Permanganate

Iron. Many substances can be determined by titration with permanganate in acid solution. The determination of iron in iron ores is one of the important applications of permanganate. The principal ores of iron are the oxides or hydrated oxides: hematite, Fe_2O_3; magnetite, Fe_3O_4; goethite, $Fe_2O_3 \cdot H_2O$; and limonite, $2Fe_2O_3 \cdot 3H_2O$. The carbonate, $FeCO_3$ (siderite), and the sulfide, FeS_2 (pyrite), are of lesser importance. The best acid for dissolving these ores is hydrochloric. The hydrated oxides dissolve readily, whereas magnetite and hematite dissolve rather slowly. The addition of tin(II) chloride aids in dissolving the latter oxides. The residue of silica remaining after the sample has been heated with acid may retain some iron. The silica can be fused with sodium carbonate and then treated with hydrochloric acid to recover the iron.

Before titrating with permanganate (or other oxidizing agent), all the iron must be reduced to the $+2$ state. There are several methods available for this reduction. We shall discuss here only the use of tin(II) chloride and amalgamated zinc (the Jones reductor).

The tin(II) ion is a stronger reducing agent ($E^\circ = +0.15$ V) than is iron(II) ion ($E^\circ = +0.77$ V). Therefore, tin(II) chloride can be used to reduce iron(III) ion to iron(II) prior to titration. Tin(II) chloride is added to the hot solution and the progress of the reduction is observed by noting the disappearance of the yellow color of iron(III) ion:

$$Sn^{2+} + 2Fe^{3+} \longrightarrow Sn^{4+} + 2Fe^{2+}$$

A slight excess of tin(II) chloride is added to ensure completeness of reduction. This excess must be removed or it will react with the oxidizing agent upon titration. For this purpose, the solution is cooled and $HgCl_2$ is added rapidly to oxidize excess Sn^{2+}:

$$2HgCl_2 + SnCl_2 \longrightarrow Hg_2Cl_2 \downarrow + SnCl_4$$

Iron(II) ion is not oxidized by $HgCl_2$. The precipitate of Hg_2Cl_2, if small, does not interfere in the subsequent titration. However, if too large an excess of $SnCl_2$ is added, Hg_2Cl_2 may be further reduced to free mercury:

$$Hg_2Cl_2 + Sn^{2+} \longrightarrow 2Hg \downarrow + 2Cl^- + Sn^{4+}$$

Mercury, which is produced in a finely divided state under these conditions, causes the precipitate to appear gray to black. If the precipitate is dark, the sample should be discarded since mercury, in the finely divided state, will be oxidized during the titration. The tendency toward further reduction of Hg_2Cl_2 is diminished if the solution is cool and $HgCl_2$ is added rapidly. Of course, if insufficient $SnCl_2$ has been added, no precipitate of Hg_2Cl_2 will be obtained. In such a case the sample must be discarded.

Tin(II) chloride is normally used for reduction of iron in samples which have been dissolved in hydrochloric acid. Zimmermann-Reinhardt preventive solution is then added if the titration is to be done with permanganate. If the acid present is sulfuric rather than hydrochloric, the Jones reductor (below) is preferred for reduction since no chloride ion is introduced in the reduction process.

Acidic solutions of Fe^{3+} can be reduced quantitatively to Fe^{2+} by metals such as zinc, cadmium, lead, or aluminum. Granulated zinc, the surface of which is amalgamated, is most frequently employed for this purpose. The solution to be reduced is poured through a column containing the amalgam and caught in the titration flask (see Fig.4.1). The metal can be used in the form of a wire spiral and dipped into the solution to be reduced. However, this does not ensure as thorough contact as does the use of a column. The reaction is

$$Zn + 2Fe^{3+} \longrightarrow 2Fe^{2+} + Zn^{2+}$$

The purpose of amalgamating the zinc is to prevent large amounts of this metal from dissolving in the acid. To form this amalgam, granulated zinc is treated with a dilute solution of a mercury(II) salt, and mercury is displaced, forming a coating of amalgam on the surface:

$$Zn + Hg^{2+} \longrightarrow Zn^{2+} + Hg$$

Hydrogen is not as easily displaced by zinc on the amalgamated surface, because of the high overvoltage of hydrogen on mercury. The principal reaction is, then, merely reduction of Fe^{3+} iron to Fe^{2+}. As noted above, the Jones reductor is ordinarily not used when the sample contains hydrochloric acid. Under these conditions considerable dissolution of zinc occurs, even though the zinc is amalgamated.

Other Reducing Agents. Many reducing agents in addition to iron(II) ion can be determined by titration with permanganate in acid solution. As mentioned in the discussion of standardization processes, arsenic can be titrated in hydrochloric acid. Antimony can also be determined, as well as nitrites (oxidized to

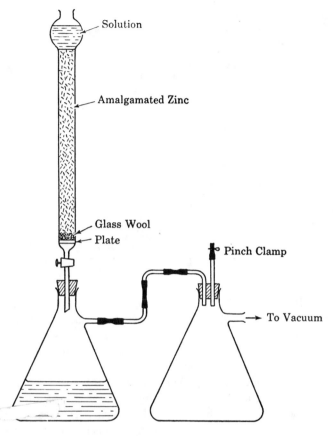

Solution

Amalgamated Zinc

Glass Wool
Plate

Pinch Clamp

To Vacuum

Fig. 4.1 Jones reductor.

nitrates) and hydrogen peroxide. The peroxide acts as a reducing agent in the reaction

$$2MnO_4^- + 5H_2O_2 + 6H^+ \longrightarrow 2Mn^{2+} + 5O_2 + 8H_2O$$

We have also seen that oxalates can be titrated with permanganate. The volumetric determination of calcium in limestone is frequently used as a student exercise. Calcium is precipitated as the oxalate. After filtration and washing, the precipitate is dissolved in sulfuric acid, and the oxalate is titrated with permanganate. The procedure is more rapid than igniting the precipitate and weighing. However, care must be taken in the precipitation to avoid contamination of the precipitate with other oxalates or oxalic acid. Such contamination, of course, leads to high results. Other cations which form insoluble oxalates can be determined in the same manner. These include manganese(II), zinc(II), cobalt(II), barium(II), strontium(II), and lead(II).

Table 4.1 summarizes some of the more common determinations that can be made by direct titration with permanganate in acid solution.

TABLE 4.1 SOME APPLICATIONS OF DIRECT
TITRATIONS WITH PERMANGANATE
IN ACID SOLUTION

Analyte	Half-Reaction of Substance Oxidized
Antimony(III)	$HSbO_2 + 2H_2O \rightleftharpoons H_3SbO_4 + 2H^+ + 2e$
Arsenic(III)	$HAsO_2 + 2H_2O \rightleftharpoons H_3AsO_4 + 2H^+ + 2e$
Bromine	$2Br^- \rightleftharpoons Br_2 + 2e$
Hydrogen peroxide	$H_2O_2 \rightleftharpoons O_2 + 2H^+ + 2e$
Iron(II)	$Fe^{2+} \rightleftharpoons Fe^{3+} + e$
Molybdenum(III)	$Mo^{3+} + 2H_2O \rightleftharpoons MoO_2^{2+} + 4H^+ + 3e$
Nitrite	$HNO_2 + H_2O \rightleftharpoons NO_3^- + 3H^+ + 2e$
Oxalate	$H_2C_2O_4 \rightleftharpoons 2CO_2 + 2H^+ + 2e$
Tin(II)	$Sn^{2+} \rightleftharpoons Sn^{4+} + 2e$
Titanium(III)	$Ti^{3+} + H_2O \rightleftharpoons TiO^{2+} + 2H^+ + e$
Tungsten(III)	$W^{3+} + 2H_2O \rightleftharpoons WO_2^{2'} + 4H^+ + 3e$
Uranium(IV)	$U^{4+} + 2H_2O \rightleftharpoons UO_2^{2+} + 4H^+ + 2e$
Vanadium(IV)	$VO^{2+} + 3H_2O \rightleftharpoons V(OH)_4^+ + 2H^+ + e$

Oxidizing Agents.   Permanganate can also be employed indirectly in the determination of oxidizing agents, particularly the higher oxides of such metals as lead and manganese. Such oxides are difficult to dissolve in acids or bases without reduction of the metal. It is impractical to titrate these substances directly because the reaction of the solid with the reducing agent is slow. Hence the sample is treated with a known excess of some reducing agent and heated to complete the reaction. Then the excess reducing agent is titrated with permanganate. Various reducing agents may be employed, such as iron(II) sulfate, arsenic(III) oxide, and sodium oxalate. The analysis of pyrolusite (MnO_2) in this manner is a common student exercise. The reaction of pyrolusite with arsenious acid is given as

$$MnO_2 + H_3AsO_3 + 2H^+ \longrightarrow Mn^{2+} + H_3AsO_4 + 2H_2O$$

EXPERIMENT 4.3.   **Determination of an Oxalate**

Procedure

Weigh accurately three portions of the dried material of appropriate size (Note). If the Fowler-Bright procedure is followed, dissolve each sample in 250 ml of dilute sulfuric acid in a 500-ml Erlenmeyer flask. The acid (12.5 ml of concentrated acid diluted to 250 ml with water) should have been previously boiled and then cooled to 24 to 30°C. Tirate as directed on page 61.

If the McBride procedure is used, dissolve each sample in 75 ml of 1.5 N sulfuric acid (20 ml of concentrated acid to 400 ml of water) in a 250-ml Erlenmeyer flask. Heat the solution almost to boiling and titrate as directed on page 62.

Calculate and report the percentage of sodium oxalate in the sample in the usual manner.

Note

Consult the instructor. The equivalent weight of sodium oxalate is 67.00. For material containing about 25% sodium oxalate, 1-g sample is a convenient amount.

EXPERIMENT 4.4. Determination of Hydrogen Peroxide

Procedure

Pipet 25 ml of the peroxide solution (Note 1) into a 250-ml volumetric flask, dilute the solution to the mark, and mix thoroughly. Transfer a 24-ml aliquot of this solution to a 250-ml Erlenmeyer flask to which has been added 5 ml of concentrated sulfuric acid and 75 ml of water. Titrate with standard permanganate solution to the first appearance of a permanent pink tinge.

Repeat the titration on two additional aliquots of the solution. Calculate the weight of hydrogen peroxide in the original 25-ml sample, and assuming that the weight was exactly 25 g, calculate the percentage by weight of hydrogen peroxide (Note 2).

Notes

1. If the density of the solution is about 1, this volume gives $25 \times 0.03 = 0.750$ g, or 750 mg of H_2O_2. Since the eq wt of H_2O_2 is 17.01, this is about 44 meq. Thus one-tenth of this solution will use about 44 ml of a 0.1 N permanganate solution for titration.

2. Commercial hydrogen peroxide often contains small amounts of organic compounds such as acetanilide, which are added to stabilize the peroxide. These compounds may react with permanganate to some extent causing incorrect results.

EXPERIMENT 4.5. Determination of Iron in an Ore

Procedure

Dissolving Sample. Weigh three samples of iron ore (Notes 1 and 2) of appropriate size (consult the instructor) into three 150-ml beakers. Add 10 ml of concentrated hydrochloric acid and 10 ml of water to each beaker. Cover the beakers with watch glasses and keep the solution just below the boiling point on a steam bath, hot plate, or wire gauze until the ore dissolves (hood) (Note 3). This may require 30 to 60 min. At this point the only solid present should be a white residue of silica. If an appreciable amount of colored solid remains in the beaker, the sample must be fused to bring the remainder of the iron into solution (Note 4). Reduce the sample according to (a) or (b) below.

Notes

1. Some commercial samples are made from iron oxide and are easily soluble in acid. Such samples will not require lengthy heating to effect solution and will not leave a residue of silica. The instructor will alter the directions if such samples are used.

2. If the ore is not finely ground, it should be ground in an agate mortar before drying. If it is suspected that the ore contains organic matter, the sample, after weighing, should be ignited in an uncovered porcelain crucible for 5 min. This oxidizes organic matter.

3. If tin(II) chloride is to be used to reduce the iron, add successive portions of stannous chloride until the solution changes from yellow to colorless. This will aid in dissolving the sample. Avoid an excess of tin(II) chloride.

4. If fusion is required (consult the instructor), dilute the solution with an equal volume of water and then filter. Wash the residue with 1% hydrochloric acid and then wash with water to remove the acid. Transfer the paper to a porcelain crucible and burn off the carbon. If the residue is white it may be disregarded, as the color was probably caused by organic matter. If the color remains, add about 5 g of potassium pyrosulfate and heat carefully until the salt just fuses. Maintain this temperature for 15 to 20 min or until all the iron reacts. Then cool and dissolve the residue in 25 ml of 1:1 hydrochloric acid and add this solution to the main filtrate.

(a) Reduction with Tin(II) Chloride. Adjust the volume of the solution to 15 to 20 ml by evaporation or dilution. The solution should be yellow in color because of the presence of iron(III) ion (Note 1). Keep the solution hot and reduce the iron in the first sample (Note 2) by adding $SnCl_2$ (Note 3) drop by drop, until the color of the solution changes from yellow to colorless (or very light green). Add 1 or 2 drops excess $SnCl_2$. Cool the solution under the tap and rapidly pour in 20 ml of saturated $HgCl_2$ solution (Note 4). Allow the solution to stand for 3 min and then rinse the solution into a 500-ml Erlenmeyer flask. Dilute to a volume of about 300 ml and add 25 ml of Zimmermann-Reinhardt solution (Note 5). Titrate slowly with permanganate, swirling the flask constantly. The end point is marked by the first appearance of a faint pink tinge which persists when the solution is swirled (Note 6).

Reduce and titrate the second and third samples in the same manner (Note 7). Calculate and report the percentage of iron in the sample in the usual manner.

Notes

1. If $SnCl_2$ has been used during dissolving, and the solution is colorless or almost so at this point, add a small crystal of potassium permanganate and heat a little longer until the yellow color is distinct. The reduction can then be followed more readily.

2. Reduce only one sample and finish the titration before reducing the second. Why?

3. This is prepared by dissolving 113 g of $SnCl_2 \cdot 2H_2O$ (free of iron) in 250 ml of concentrated hydrochloric acid, adding a few pieces of mossy tin, and diluting to 1 liter with water.

4. If the HgCl$_2$ were added slowly, part of it would be temporarily in contact with an excess of SnCl$_2$, which might reduce the substance to metallic mercury. Also, if the solution were hot, there would be a danger of forming mercury. The precipitate here should be white and silky and not large in quantity. If the precipitate is gray or black, indicating the presence of mercury, the sample should be discarded. If no precipitate is obtained, indicating insufficient SnCl$_2$, the sample should be discarded.

5. This is prepared as follows: Dissolve 70 g of MnSO$_4$ in 500 ml of water and add slowly, with stirring, 110 ml of concentrated sulfuric acid and 200 ml of 85% phosphoric acid. Dilute to 1 liter.

6. The color may fade slowly because of oxidation of Hg$_2$Cl$_2$ or chloride ion by permanganate.

7. If desired (consult the instructor), a blank can be determined by carrying a mixture of 10 ml of concentrated hydrochloric acid and 10 ml of water through the entire procedure. The blank normally will be about 0.03 to 0.05 ml of 0.1 N potassium permanganate.

(b) Jones Reductor. Add 5 ml of concentrated sulfuric acid to each beaker and evaporate the solution carefully until heavy white fumes appear (hood). This expels hydrochloric acid. Cool the solution and dilute to about 100 ml.

Arrange a Jones reductor as shown in Fig. 4.1. The column is about 2 cm inside diameter and about 45 cm long. In the lower end of the column is a perforated porcelain plate covered with glass wool. The column is filled with amalgamated zinc (Note 1), the zinc column being about 35 to 40 cm long. The tube below the stopcock is connected through a rubber stopper to a 250-ml suction flask. The side arm of the suction flask is connected to a second suction flask (as shown in Fig. 4.1), which acts as a water trap, and suction is applied to the arm of this second flask.

Wash the zinc column first with several portions of water, and then with 200 to 250 ml of 1 N sulfuric acid (about 30 ml of concentrated acid to 1 liter of water). The rate of flow of the solution through the column should be about 50 to 75 ml/min, and the liquid level should always be left just above the top of the zinc column (Note 2). Add to the solution collected in the receiving flask (about 250 ml) 0.1 N potassium permanganate solution dropwise (Note 3). If no more than 2 drops are required to impart a faint pink tinge to the solution, the reductor is ready for use. Otherwise, continue the washing with 250-ml portions of acid until no more than 2 drops of permanganate are required. Once a satisfactory blank has been obtained, place a clean receiving flask under the column and proceed to reduce the first sample.

Carefully pour the first sample through the reductor, controlling the rate of flow with the stopcock to about 50 ml/min. Stop the flow of the column when the solution stands just above the top of the zinc. Rinse the beaker that contained the sample with five 10-ml portions of 1 N sulfuric acid, and pour these washings through the column at the same rate of flow as before. Then wash the column with 100 ml of water (Note 4).

Remove the receiving flask from the column, rinse the end of the delivery tube, and add 5 ml of 85% phosphoric acid to the flask. Titrate immediately with standard permanganate, adding the solution rapidly at first, but dropwise near the end point, so that the end point will not be overstepped.

Reduce and titrate the other two samples in the same manner. Calculate the percentage of iron in the sample, remembering to subtract from the total volume of permanganate the volume used by the blank. The average deviation should be about 1 to 2 parts per thousand.

Notes

1. Prepare the amalgamated zinc as follows: To 300 ml of pure 20- to 30-mesh zinc in a beaker, add about 300 ml of a 2% mercury(II) nitrate (or chloride solution, acidified with 2 ml of nitric acid. Stir for 5 min and then wash the zinc several times by decantation. The zinc should now have a bright silvery luster from the coating of mercury. Fill the reductor tube with water and then add the zinc slowly to the column.

2. From this point on, do not allow the top of the column to be exposed to the air. Hydrogen peroxide may be formed by the reaction of atmospheric oxygen and hydrogen liberated in the reductor. Permanganate oxidizes hydrogen peroxide, and hence this effect will cause high results for iron.

3. The blank is necessary because the zinc may contain materials, particularly iron, that are oxidized by permanganate. New reductors frequently require three or four treatments before giving a satisfactory blank. If additional treatments are indicated, replace the zinc with a better grade.

4. The reductor should be left filled with water after use to prevent formation of basic salts, which tend to clog the column. No acid should be left in the column since it will eventually dissolve the zinc. For storage, the amalgam can be washed, transferred to a stoppered bottle, and covered with water.

EXPERIMENT 4.6.   Determination of Oxygen in Pyrolusite

Procedure

Weigh accurately three samples of about 0.5 g of the finely ground and dried ore into 250-ml Erlenmeyer flasks. Calculate approximately the number of milli-equivalents of oxygen in the sample (Note 1) and then the weight of sodium oxalate required to react with the sample. Add to each flask about 0.25 g of pure sodium oxalate in excess of the calculated amount, weighing this accurately on the balance. Add to each flask about 100 ml of sulfuric acid (10 ml of concentrated acid to 100 ml of water), cover the flask with a watch glass, and heat each flask on a steam bath or gently over a low flame. Shake the flask occasionally. Keep the temperature just below the boiling point and avoid allowing the solution to evaporate very much. After the samples have completely digested, as indicated by the disappearance of all the black or brown particles (Note 2) and the cessation of evolution of carbon dioxide, rinse the watch glass and rinse down the walls of the flask with distilled water. Titrate the hot solution with standard permanganate to the first appearance of a faint pink tinge.

After digestion and titration of all three samples, calculate the percentage of oxygen in the sample. The average deviation may be as high as 4 parts per thousand (consult the instructor). Remember that the milliequivalents of oxygen are given by the relationship

$$\text{meq oxygen} + \text{meq KMnO}_4 = \text{meq Na}_2\text{C}_2\text{O}_4$$

Notes

1. Consult the instructor for approximate percentage of oxygen in the sample. For example, if the sample is about 10% oxygen, a 500-mg sample would contain 50 mg of oxygen, or about 6 meq. Six milliequivalents of sodium oxalate (eq wt of 67) weigh about 0.4 g. Hence one should take about 0.65 g of sodium oxalate.

2. There may be a residue of white or light-brownish silica. This should be disregarded.

COMPOUNDS OF CERIUM

Properties

The element cerium (atomic number 58) can exist in only two oxidation states, $+4$ and $+3$. In the quadrivalent state it is a powerful oxidizing agent, undergoing a single reaction

$$\text{Ce}^{4+} + e \rightleftharpoons \text{Ce}^{3+}$$

The Ce(IV) ion is used in solutions of high acidity because hydrolysis leads to precipitation in solutions of low hydrogen ion concentration. The redox potential of the Ce(IV)–Ce(III) couple is dependent upon the nature and concentration of the acid present. The formal potentials in 1 M solutions of the common acids are: HClO_4, $+1.70$ V; HNO_3, $+1.61$ V; H_2SO_4, $+1.44$ V; HCl, $+1.28$ V.

It is known that both cerium(IV) and cerium(III) ions form stable complexes with various anions. Some chemists name the acids and salts of cerium to indicate that the element is present as a complex anion, rather than as a cation. For example, the salt $(\text{NH}_4)_2\text{Ce}(\text{NO}_3)_6$ is called *ammonium hexanitratocerate*. For simplicity, we shall call such a compound cerium(IV) ammonium nitrate and write the formula $\text{Ce}(\text{NO}_3)_4 \cdot 2\text{NH}_4\text{NO}_3$.

Although cerium is a rare earth element, its compounds are readily available for analytical use at a reasonable cost. Since 1928, beginning with the work of N. H. Furman at Princeton and H. H. Willard at Michigan, the reagent has found ever-increasing use as an oxidizing agent in analytical chemistry.[6] It is usually

---

[6] See the booklet entitled *Cerate Oxidimetry*, published by the G. Frederick Smith Chemical Co., Columbus, Ohio, 1942, and similar booklets; see also I. M. Kolthoff and R. Belcher, *Volumetric Analysis*, Vol. 3, Interscience Publishers, Inc., New York, 1957, p. 121.

necessary to employ a redox indicator and the compound *ferroin* (Chapter 10 of text) has been developed for this purpose.

The Ce(IV) ion can be employed in most titrations where permanganate is used, and it possesses properties which often make it a better choice as an oxidizing agent than permanganate. The principal advantages can be summarized as follows:

1. There is only one oxidation state, Ce(III) to which the Ce(IV) ion is reduced.
2. It is a very strong oxidizing agent, and, as pointed out above, one can vary the intensity of its oxidizing power by choice of the acid employed.
3. Sulfuric acid solutions of Ce(IV) ion are extremely stable. Solutions can be kept indefinitely without change in concentration. Solutions in nitric and perchloric acids decompose, but only slowly.
4. The chloride ion, in moderate concentration, is not readily oxidized, even in the presence of iron. The reagent can thus be used for iron titrations in hydrochloric acid solution without need of Zimmermann-Reinhardt preventive solution. Cerium(IV) solutions can be employed, even in the presence of chloride ion, for oxidations which must be carried out by use of excess reagent at elevated temperature. However, chloride ion is oxidized if the solution is boiled. Also, Ce(IV) solutions in hydrochloric acid are unstable if the concentration of acid is more than about 1 M.
5. A salt, cerium(IV) ammonium nitrate, sufficiently pure to be weighed directly for preparing standard solutions, is available.
6. Although the Ce(IV) ion is yellow, the color does not cause difficulty in reading a buret, unless the concentration is greater than about 0.1 M. The Ce(III) ion is colorless.

Standardization of Solutions

Solutions of Ce(IV) are usually prepared from cerium(IV) hydrogen sulfate, $Ce(HSO_4)_4$, cerium(IV) ammonium sulfate, $Ce(SO_4)_2 \cdot 2(NH_4)_2SO_4 \cdot 2H_2O$, or cerium(IV) hydroxide, $Ce(OH)_4$. The compounds are dissolved in 0.2 to 0.5 M strong acid to prevent hydrolysis and the formation of slightly soluble basic salts. The solution is then standardized against one of the primary standards listed below.

As mentioned above, the compound cerium(IV) ammonium nitrate is available as a primary standard and standard solutions may be made by direct weighing, followed by dilution in a volumetric flask. This salt can also be obtained as ordinary reagent-grade material, in which case the solution must be standardized.

The same primary standards used for potassium permanganate can be used to standardize Ce(IV) solutions. These are:

1. *Arsenic(III) oxide.* The reaction is slow unless osmium tetroxide, OsO_4, or iodine monochloride, ICl, is used as a catalyst. With OsO_4 catalyst the titration can be made at room temperature. Ferroin is a suitable indicator.

flask. Dilute to the mark and mix the solution thoroughly. Calculate the normality of the solution.

If the solution is to be standardized against iron (consult the instructor), proceed as follows. Weigh and dissolve samples of iron wire as directed under Experiment 4.8, page 75. Reduce the Fe^{3+} with $SnCl_2$ as directed there, and rinse the solution into a 500-ml Erlenmeyer flask. Now add 250 ml of water, 5 ml of concentrated sulfuric acid, 5 ml of 85% phosphoric acid, and 8 drops of sodium diphenylaminesulfonate indicator (Note 2). Titrate slowly with the dichromate solution, swirling the flask constantly. The solution becomes green since the reaction produces Cr^{3+} ions. The oxidized form of the indicator is purple or violet. As the end point is approached, indicated by a transient purple color, add the dichromate dropwise until the color changes permanently to purple.

Calculate the normality of the dichromate solution. Compare this with the normality calculated from the weight of dichromate dissolved in the solution.

Notes

1. The equivalent weight is 294.18/6 or 49.03.
2. The indicator solution is made by dissolving 0.3 g of the sodium salt in 100 ml of water. If the barium salt is used, dissolve 0.3 g in 100 ml of hot water, add an excess of 0.1 M sodium sulfate solution, and let the solution stand until barium sulfate settles. Either filter or decant the solution from the precipitate.

EXPERIMENT 4.11. **Determination of Iron in an Ore, Using Dichromate**

Procedure

Weigh and dissolve the first sample and reduce it with $SnCl_2$ as directed under Experiment 4.5. Then cool the solution and rapidly add 20 ml of saturated $HgCl_2$ solution. Allow the solution to stand for 3 min and then rinse into a 500-ml Erlenmeyer flask. Now add 250 ml of water, 5 ml of concentrated sulfuric acid, 5 ml of 85% phosphoric acid, and 8 drops of sodium diphenylaminesulfonate indicator (see Note 2, Experiment 4.10). Titrate with dichromate solution as directed in Experiment 4.10.

Reduce and titrate the other two samples in the same manner. Calculate and report the percentage of iron in the ore.

IODINE

The iodine–iodide redox system,[8]

---

[8] The principal species in a solution of iodine and potassium iodide is the triiodide ion, I_3^-, and many chemists refer to *triiodide* solutions rather than *iodine* solutions. For simplicity, we shall continue to use the term *iodine* solutions and write equations using I_2 rather than I_3^-.

$$I_3^- + 2e \rightleftharpoons 3I^-$$

has a standard potential of $+0.53$ V. Iodine, therefore, is a much weaker oxidizing agent than are potassium permanganate, cerium (IV) compounds, and potassium dichromate. On the other hand, iodide ion is a reasonably strong reducing agent—stronger, for example, than iron(II) ion. In actual analytical processes, iodine is used as an oxidizing agent (*iodimetry*) and iodide ion is used as a reducing agent (*iodometry*). Relatively few substances are sufficiently strong reducing agents to be titrated directly with iodine. Hence the number of iodimetric determinations is small. However, many oxidizing agents are sufficiently strong to react completely with iodide ion, and there are many applications of iodometric processes. An excess of iodide ion is added to the oxidizing agent being determined, liberating iodine, which is then titrated with sodium thiosulfate solution. The reaction between iodine and thiosulfate goes well to completion. This reaction is discussed below. It should be pointed out that some authors prefer to avoid the term *iodimetry*, and instead, speak of direct and indirect *iodometric* processes.

Direct, or Iodimetric Processes

The more important substances that are sufficiently strong reducing agents to be titrated directly with iodine are thiosulfate, arsenic(III), antimony(III), sulfide, sulfite, tin(II), and ferrocyanide ions. The reducing power of several of these substances depends upon the hydrogen ion concentration, and only by proper adjustment of pH can the reaction with iodine be made quantitative. Before discussing such reactions, let us consider the preparation and properties of iodine solutions.

Preparation of Solution. Iodine is only slightly soluble in water (0.00134 mol/liter at 25°C), but is quite soluble in solutions containing iodide ion. Iodine forms a complex ion with iodide,

$$I_2 + I^- \rightleftharpoons I_3^-$$

the equilibrium constant[9] being about 710 at 25°C. An excess of potassium iodide is added to increase the solubility and to decrease the volatility of iodine. Usually about 3 to 4% by weight of potassium iodide is added to a 0.1 N solution, and the bottle containing the solution is well stoppered.

Iodine tends to hydrolyze in water, forming hydriodic and hypoiodous acids,

$$I_2 + H_2O \longrightarrow HI + HIO$$

Conditions which increase the degree of hydrolysis must be avoided. Titrations cannot be made in very basic solutions, and standard solutions of iodine must be kept in dark bottles to prevent the decomposition of hypoiodous acid by sunlight,

[9] W. M. Latimer, *Oxidation Potentials*, Prentice-Hall, Inc., Englewood Cliffs, N.J., 1952.

$$2HIO \longrightarrow 2HI + O_2$$

Hypoiodous acid may also be converted into iodate in basic solution,

$$3HIO + 3OH^- \longrightarrow 2I^- + IO_3^- + 3H_2O$$

Standardization. Standard iodine solutions can be prepared by direct weighing of pure iodine and dilution in a volumetric flask. The iodine is purified by sublimation and is added to a concentrated potassium iodide solution which is weighed accurately before and after the addition of iodine. Usually, however, the solution is standardized against a primary standard, As_2O_3 being most commonly used. The reducing power of $HAsO_2$ depends upon the pH, as shown by the following reaction:

$$HAsO_2 + I_2 + 2H_2O \rightleftharpoons H_3AsO_4 + 2I^- + 2H^+$$

The experimental value of the equilibrium constant of this reaction is 0.17, the equilibrium lying more to the left than to the right. However, by lowering the hydrogen ion concentration, the reaction is forced to the right and can be made sufficiently complete to be suitable for a titration.

From equilibrium considerations, pH values between about 5 and 11 are permissible for the titration of $HAsO_2$ with iodine.[10] However, a pH of about 8 is usually employed for the titration for the following reasons. The rate of reaction between iodine and $HAsO_2$ is slow if the pH is less than about 7. If the pH is much greater than 9, hydrolysis and subsequent formation of IO_3^- may occur locally where drops of titrant strike the solution. Sodium bicarbonate is usually added to the $HAsO_2$ solution before titration with iodine. The solution is then buffered at a pH slightly over 8 and the titration gives excellent results.

Starch Indicator. The color of a 0.1 N solution of iodine is sufficiently intense so that iodine can act as its own indicator. Iodine also imparts an intense purple or violet color to such solvents as carbon tetrachloride or chloroform, and sometimes this is utilized in detecting the end point of titrations. More commonly, however, a solution (colloidal dispersion) of starch is employed since the deep blue color of the starch-iodine complex serves as a very sensitive test for iodine. The sensitivity is greater in slightly acid than in neutral solutions and is greater in the presence of iodide ions.

The exact mechanism of the formation of the colored complex between starch and iodine is not known. However, it is thought that molecules of iodine are held on the surface of β-amylose, a constituent of starch.[11] Another constituent of starch,

[10] E. W. Washburn, *J. Am. Chem. Soc.*, **30**, 31 (1908); R. K. McAlpine, *J. Chem. Ed.*, **26**, 362 (1949).

[11] R. E. Rundle, J. F. Foster, and R. R. Baldwin, *J. Am. Chem. Soc.*, **66**, 2116 (1944).

TABLE 4.2 DETERMINATIONS BY DIRECT IODINE TITRATIONS

Analyte	Reaction
Antimony(III)	$HSbOC_4H_6O_6 + I_2 + H_2O \rightleftharpoons HSbO_2C_4H_4O_6 + 2H^+ + 2I^-$
Arsenic(III)	$HAsO_2 + I_2 + 2H_2O \rightleftharpoons H_3AsO_4 + 2H^+ + 2I^-$
Ferrocyanide	$2Fe\,(CN)_6^{4-} + I_2 \rightleftharpoons 2Fe\,(CN)_6^{3-} + 2I^-$
Hydrogen cyanide	$HCN + I_2 \rightleftharpoons ICN + H^+ + I^-$
Hydrazine	$N_2H_4 + 2I_2 \rightleftharpoons N_2 + 4H^+ + 4I^-$
Sulfur (sulfide)	$H_2S + I_2 \rightleftharpoons 2H^+ + 2I^- + S$
Sulfur (sulfite)	$H_2SO_3 + I_2 + H_2O \rightleftharpoons H_2SO_4 + 2H^+ + 2I^-$
Thiosulfate	$2S_2O_3^{2-} + I_2 \rightleftharpoons S_4O_6^{2-} + 2I^-$
Tin(II)	$Sn^{2+} + I_2 \rightleftharpoons Sn^{4+} + 2I^-$

α-amylose or amylopectin, forms reddish complexes with iodine which are not easily decolorized. Therefore, starches that contain much amylopectin should not be used. So-called "soluble starch" is principally β-amylose.

Starch solutions are easily decomposed by bacteria, a process which may be retarded by sterilization or addition of a preservative. The decomposition products consume iodine and turn reddish. Mercury(II) iodide, boric acid, or furoic acid can be used as preservatives. Conditions that lead to hydrolysis or coagulation of starch should be avoided. The indicator sensitivity is decreased by increasing temperature and by some organic materials such as methyl and ethyl alcohols.

Determinations with Iodine. Some of the determinations that can be done by direct titration with a standard iodine solution are listed in Table 4.2. The determination of antimony is similar to that of arsenic except that tartrate ions, $C_4H_4O_6^{2-}$, are added to complex antimony and avoid precipitation of salts such as SbOCl, when the solution is neutralized. The titration is carried out in a bicarbonate buffer of *p*H about 8. In the determination of tin and sulfites the solution being titrated must be protected from oxidation by air. The hydrogen sulfide titration is frequently used to determine sulfur in iron or steel.

EXPERIMENT 4.12. **Preparation and Standardization of a 0.1 *N* Iodine Solution**

Procedure

Weigh on a trip balance about 12.7 g of reagent-grade iodine (Notes 1 and 2) and place this in a 250-ml beaker. Add to the beaker 40 g of potassium iodide, free of iodate (Note 3), and 25 ml of water. Stir to dissolve all the iodine and transfer the solution to a glass-stoppered bottle. Dilute to about 1 liter (Note 4).

Standardize the iodine solution as follows. Weigh accurately a sample of reagent-grade As_2O_3 (Note 5) of about 1.25 g into a 250-ml beaker. Add a solution made by dissolving 3 g of NaOH in 10 ml of water. Allow the beaker to stand, swirling it occasionally, until the solid has completely dissolved. Then add 50 ml of water, 2 drops of phenolphthalein indicator, and 1:1 hydrochloric acid until the pink color of the indicator just disappears. Then add 1 ml of hydrochloric acid in excess. Transfer the solution to a 250-ml volumetric flask, dilute to the mark, and mix thoroughly.

Transfer with a pipet 25 ml of the arsenic(III) solution to a 250-ml Erlenmeyer flask and dilute with 50 ml of water. Then carefully add small portions of sodium bicarbonate to neutralize the acid. When vigorous effervescence ceases, add 3 g of additional sodium bicarbonate to buffer the solution. Add 5 ml of starch indicator (Note 6) and titrate with iodine until the first appearance of the deep blue color, which persists for at least 1 min. If the end point is overrun, some of the standard arsenic(III) solution can be used for back-titration.

Titrate two other aliquots in the same manner. Calculate the normality of the arsenic(III) solution from the weight of As_2O_3 taken. Then calculate the normality of the iodine solution from the volumes of the two solutions used.

Notes

1. If the solution is to be standardized and only one unknown analyzed, 500 ml of solution will suffice. The quantities of iodine and potassium iodide can be halved (consult the instructor).

2. If desired, the iodine can be weighed accurately and the solution made up to a definite volume. To prepare 500 ml of solution, first place about 20 g of potassium iodide in a large weighing bottle and dissolve this in 10 ml of water. After the solution has come to room temperature, weigh the bottle accurately (to the nearest milligram is sufficient). Take the bottle to a trip balance and there add about 6.4 g of iodine and stopper the bottle tightly. Do not open the bottle near the analytical balance as iodine fumes are corrosive. Reweigh the bottle and record the weight of iodine taken. When the iodine has dissolved, transfer the solution to a 500-ml volumetric flask, dilute to the mark, and mix thoroughly. Store the solution in a glass-stoppered bottle.

3. Test the salt for iodate as follows: Dissolve about 1 g in 20 ml of water and add 1 ml of 6 N sulfuric acid and 2 ml of starch solution. No blue color should develop in 30 s.

4. A bottle of brown glass is preferable for storing this solution. In any event, the solution should be kept from the light as much as possible.

5. Pure As_2O_3 is not hygroscopic and may not need drying unless it has been exposed to air for some time. If drying is necessary, it is usually preferable to place the material in a desiccator over sulfuric acid for about 12 h. The oxide (octahedral variety) tends to sublime when heated as high as 125 to 150°C.

6. This solution is prepared as follows: Make a paste of 2 g of soluble starch and 25 ml of water and pour this gradually (with stirring) into 500 ml of boiling water. Continue the boiling for 1 or 2 min, add 1 g of boric acid as a preservative, and allow the solution to cool. Store the solution in a glass-stoppered bottle. Alternatively, a solid complex of starch and urea can be prepared by melting urea in a small beaker and adding soluble starch. The ratio of weights of urea to starch should be 4:1. The solution is mixed, allowed to cool and solidify, and then the solid is ground in a mortar. A small scoopful is used in place of starch solution. The complex dissolves readily in water and is quite stable.

EXPERIMENT 4.13.   **Determination of the Purity of Arsenic(III) Oxide**

Procedure

Weigh three samples of the unknown of appropriate size (Note) into 500-ml Erlenmeyer flasks. Add to each flask 50 ml of water and 1 g of sodium hydroxide and warm until the sample dissolves. Cool the solution, add 2 drops of phenolphthalein indicator, and add 1:1 hydrochloric acid until the pink color of the indicator just disappears. Then add small portions of sodium bicarbonate to neutralize the acid. When vigorous effervescence ceases, add 3 g of additional bicarbonate to buffer the solution. Dilute to about 150 ml with water and add 5 ml of starch solution (Note 6, Experiment 4.12). Titrate the first sample with iodine solution to the first appearance of the deep blue color, which lasts for at least 1 min.

Titrate the other two samples in the same manner. Calculate and report the percentage of As_2O_3 in the sample.

Note

Consult the instructor as to size of sample and method of drying. Commercial unknown samples are usually about 3 to 15% As_2O_3.

EXPERIMENT 4.14.   **Determination of Vitamin C**

Vitamin C (ascorbic acid) is a reducing agent and can be determined by titration with standard iodine solution:

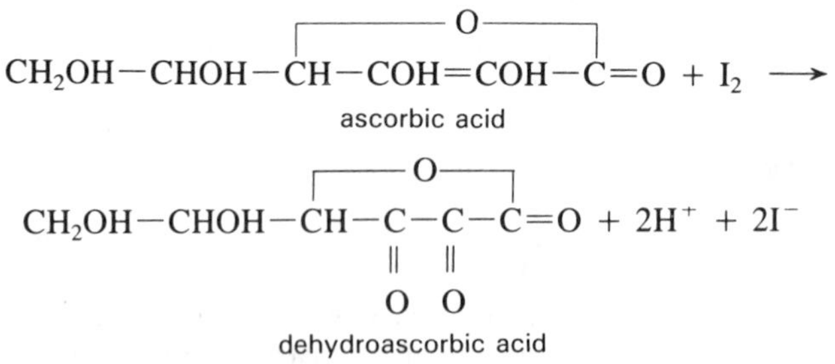

$$CH_2OH-CHOH-\overset{\displaystyle \lceil\;\;\;O\;\;\;\rceil}{CH-COH}=COH-C=O + I_2 \longrightarrow$$

ascorbic acid

$$CH_2OH-CHOH-\overset{\displaystyle \lceil\;\;O\;\;\rceil}{CH-\underset{\parallel \;\; \parallel}{C-C}-C}=O + 2H^+ + 2I^-$$
$$O\;\;O$$

dehydroascorbic acid

Since the molecule loses two electrons in this reaction its equivalent weight is one-half the molecular weight, or 88.07 g/eq.

Procedure

Weigh accurately three 100-mg vitamin C tablets (Note 1) and place them in a 250-ml Erlenmeyer flask. Dissolve the tablets in about 50 ml of water, swirling the flask to hasten dissolution of the tablets (Note 2). Add 5 ml of starch indicator

(Note 6, Experiment 4.12) and titrate immediately (Note 3) with standard I_2 solution to the first appearance of the deep blue color, which persists for at least 1 min.

Repeat the foregoing procedure twice, remembering to carry out the titration as soon as the sample is dissolved. Calculate the number of milligrams of vitamin C per tablet.

Notes

1. Three hundred milligrams of ascorbic acid should require about 34 ml of 0.1 N I_2 for titration.

2. A stirring rod can be used to break up the tablets. A small amount of binder in the tablets will not dissolve and will remain as a suspension.

3. A solution of vitamin C is readily oxidized by oxygen in the air, and the titration should be carried out as soon as the sample is dissolved. During the titration the flask can be covered by a piece of cardboard having a small hole for the tip of the buret.

Indirect, or Iodometric Processes

Many strong oxidizing agents can be analyzed by adding excess potassium iodide and titrating the liberated iodine. Many oxidizing agents require acid solutions for reaction with iodide; thus sodium arsenite, which must be used in slightly alkaline solution, is not a convenient titrant for iodine. Sodium thiosulfate, which can be used in acid solution, is therefore commonly used as the titrant.

The solution of potassium iodide need not be standardized since an excess of the salt is added. However, proper precautions must be taken in its use to avoid errors. For example, iodide ion is oxidized by air:

$$4H^+ + 4I^- + O_2 \longrightarrow 2I_2 + 2H_2O$$

This reaction is slow in neutral solution but is faster in acid and is accelerated by sunlight. After addition of potassium iodide to an acid solution of an oxidizing agent, the solution should not be allowed to stand for a long time in contact with air, since additional iodine will be liberated. Nitrites should not be present because they are reduced by iodide ion to nitric oxide, which in turn is oxidized back to nitrite by oxygen of the air:

$$2HNO_2 + 2I^- + 2H^+ \longrightarrow 2NO + I_2 + 2H_2O$$

$$4NO + O_2 + 2H_2O \longrightarrow 4HNC_z$$

The potassium iodide should be free of iodates since these two substances react in acid solution to liberate iodine:

$$IO_3^- + 5I^- + 6H^+ \longrightarrow 3I_2 + 3H_2O$$

Sodium Thiosulfate

The standard solution employed in most iodometric processes is sodium thiosulfate. This salt is commonly purchased as the pentahydrate $Na_2S_2O_3 \cdot 5H_2O$. Solutions should not be prepared by direct weighing, but should be standardized against a primary standard.

Sodium thiosulfate solutions are not stable over long periods of time. Sulfur-consuming bacteria find their way into the solution, and their metabolic processes lead to the formation of sulfite, sulfate, and colloidal sulfur. The presence of the latter causes turbidity, the appearance of which justifies discarding the solutions. Normally, the water used to prepare thiosulfate solutions is boiled to render it sterile, and frequently borax or sodium carbonate is added as a preservative. Air oxidation of thiosulfate is very slow. However, a trace of copper, which is often present in distilled water, will catalyze oxidation by air, probably reacting as

$$2Cu^{2+} + 2S_2O_3^{2-} \longrightarrow 2Cu^+ + S_4O_6^{2-}$$

and

$$2Cu^+ + \tfrac{1}{2}O_2 + 2H^+ \longrightarrow 2Cu^{2+} + H_2O$$

A basic compound such as sodium carbonate decreases the extent of the second reaction.

Thiosulfate is decomposed in acidic solutions, liberating sulfur as a milky precipitate:

$$S_2O_3^{2-} + 2H^+ \longrightarrow H_2S_2O_3 \longrightarrow H_2SO_3 + S(s)$$

This reaction is slow, however, and does not occur when thiosulfate is titrated into strongly acidic solutions of iodine if the solutions are well stirred. The reaction between iodine and thiosulfate is much more rapid than the reaction with hydrogen ions.

Iodine oxidizes thiosulfate to the tetrathionate ion:

$$I_2 + 2S_2O_3^{2-} \longrightarrow 2I^- + S_4O_6^{2-}$$

The reaction is rapid and goes well to completion. The equivalent weight of sodium thiosulfate pentahydrate, $Na_2S_2O_3 \cdot 5H_2O$, is the molecular weight 248.17, since one electron per molecule is lost. If the solution is fairly basic (above a pH of about 9), thiosulfate is oxidized partially to sulfate:

$$4I_2 + S_2O_3^{2-} + 5H_2O \longrightarrow 8I^- + 2SO_4^{2-} + 10H^+$$

Standardization of Thiosulfate Solutions

A number of substances can be employed as primary standards for thiosulfate solutions. Pure iodine is the most obvious standard, but is seldom used because of difficulty in handling and weighing. More often, use is made of a strong oxidizing agent which will liberate iodine from iodide, an iodometric process.

Potassium Dichromate. This compound can be obtained in a high degree of purity. It has a fairly high equivalent weight, it is nonhygroscopic, and the solid and its solutions are very stable. The reaction with iodide is carried out in about 0.2 to 0.4 M acid and is complete in 5 to 10 min:

$$K_2Cr_2O_7 + 6I^- + 14H^+ \longrightarrow 2K^+ + 2Cr^{3+} + 3I_2 + 7H_2O$$

The equivalent weight of potassium dichromate is one-sixth of the molecular weight, or 49.03. At acid concentrations greater than 0.4 M, air oxidation of potassium iodide becomes appreciable. For best results, a small portion of sodium bicarbonate or dry ice is added to the titration flask. The carbon dioxide produced displaces the air, after which the mixture is allowed to stand until the reaction is complete.

Potassium Iodate and Potassium Bromate. Both of these salts oxidize iodide quantitatively to iodine in acid solution:

$$IO_3^- + 5I^- + 6H^+ \longrightarrow 3I_2 + 3H_2O$$

$$BrO_3^- + 6I^- + 6H^+ \longrightarrow 3I_2 + Br^- + 3H_2O$$

The iodate reaction is quite rapid; it also requires only a slight excess of hydrogen ions for complete reaction. The bromate reaction is rather slow, but the speed can be increased by increasing the hydrogen ion concentration. Usually, a small amount of ammonium molybdate is added as a catalyst.

The principal disadvantage of these two salts as primary standards is that the equivalent weights are small. In each case the equivalent weight is one-sixth of the molecular weight,[12] that of potassium iodate being 35.67 and that of potassium bromate being 27.84. In order to avoid a large error in weighing, directions usually call for weighing a large sample, diluting in a volumetric flask, and withdrawing aliquot portions. Both salts can be obtained sufficiently pure for use as standards, although for very accurate work potassium iodate is not recommended. The salt, potassium acid iodate, $KIO_3 \cdot HIO_3$, can also be used as a primary standard, but its equivalent weight is small, one-twelfth the molecular weight, or 32.49.

---

[12] It should be noted that the iodate ion gains five electrons in the reaction with iodide ions, and therefore its equivalent weight *in this reaction* is one-fifth of the molecular weight. However, the reaction involved in the titration is that between iodine and sodium thiosulfate. Since 1 mmol of iodate produces 3 mmol, or 6 meq, of iodine, the equivalent weight of iodate for the complete process is one-sixth of the molecular weight.

Copper. Pure copper can be used as a primary standard for sodium thiosulfate and is recommended when the thiosulfate is to be used for the determination of copper. The redox potential of the cuprous-cupric system,

$$Cu^{2+} + e \rightleftharpoons Cu^+$$

is $+0.15$ V, and thus iodine ($E° = +0.53$ V) is a better oxidizing agent than Cu(II) ion. However, when iodide ions are added to Cu(II) ions, a precipitate of CuI is formed:

$$2Cu^{2+} + 4I^- \longrightarrow 2CuI(s) + I_2$$

The reaction is forced to the right by removal of Cu(I) ion and also by addition of excess iodide.

The *p*H of the solution must be maintained by a buffer system, preferably between 3 and 4. At higher *p*H values partial hydrolysis of Cu(II) ion takes place, and the reaction with iodide is slow. In strong acid solutions, the copper-catalyzed oxidation of iodide by air is appreciable.

If the anion (such as acetate ion) used in the buffer forms a fairly stable complex with Cu(II) ion, the reaction between Cu(II) and iodide ions may be prevented from going to completion. As iodine is removed by titration with thiosulfate, the Cu(II) complex dissociates to form more Cu(II) ions, which in turn react with iodide ions to liberate more iodine. This results in a recurring end point.

It has been found that iodine (I_3^- ion) is held by adsorption on the surface of the CuI precipitate, rendering it grayish rather than white. Unless the iodine is displaced, the end point is reached too soon and may recur if iodine is slowly desorbed from the surface. Foote and Vance[13] found that addition of potassium thiocyanate, just before the end point is reached, gives a sharper color change. There are two reasons for this. First, copper(I) thiocyanate is less soluble than CuI; thus thiocyanate aids in forcing the reaction to completion, even in the presence of complexing anions:

$$2Cu^{2+} + 2I^- + 2SCN^- \longrightarrow 2CuSCN(s) + I_2$$

Second, CuSCN may be formed on the surface of CuI particles already precipitated:

$$CuI + SCN^- \longrightarrow CuSCN + I^-$$

Thiocyanate ion is more strongly adsorbed than is I_3^- ion on the CuSCN surface (Paneth-Fajans-Hahn rule). In other words, adsorbed iodine is displaced from the surface and hence is able to react rapidly with thiosulfate. The thiocyanate should not be added until most of the iodine has been titrated, since iodine may oxidize thiocyanate.

---

[13]H. W. Foote and J. E. Vance, *J. Am. Chem. Soc.*, **57**, 845 (1935); *Ind. Eng. Chem., Anal. Ed.*, **8**, 119 (1936).

Iodometric Determinations

There are many applications of iodometric processes in analytical chemistry. Some of these are listed in Table 4.3. The iodometric determination of copper is widely used for both ores and alloys. The method gives excellent results and is more rapid than the electrolytic determination of copper. Copper ores commonly contain iron, arsenic, and antimony. These elements in their higher oxidation states (usually so from the dissolving process) will oxidize iodide and thus interfere. The interference of iron can be prevented by addition of ammonium bifluoride, NH_4HF_2, which converts iron(III) ion into the stable complex,[14] FeF_6^{3-}. As previously mentioned, antimony and arsenic will not oxidize iodide ions except in solutions of high acidity. By adjusting the pH to about 3.5 with a buffer, interference from these two elements is eliminated. Park[15] suggested using a phthalate buffer for this purpose. However, later investigations[16] showed that a solution of bifluoride ion, HF_2^-, added to complex iron, gives a buffer of approximately the desired pH, so that no additional buffer is needed.

The classical method of Winkler[17] is a sensitive method for determining

TABLE 4.3 DETERMINATIONS BY INDIRECT IODINE TITRATIONS

Analyte	Reaction
Arsenic(V)	$H_3AsO_4 + 2H^+ + 2I^- \rightleftharpoons HAsO_2 + I_2 + 2H_2O$
Bromine	$Br_2 + 2I^- \rightleftharpoons 2Br^- + I_2$
Bromate	$BrO_3^- + 6H^+ + 6I^- \rightleftharpoons Br^- + 3I_2 + 3H_2O$
Chlorine	$Cl_2 + 2I^- \rightleftharpoons 2Cl^- + I_2$
Chlorate	$ClO_3^- + 6H^+ + 6I^- \rightleftharpoons Cl^- + 3I_2 + 3H_2O$
Copper(II)	$2Cu^{2+} + 4I^- \rightleftharpoons 2CuI(s) + I_2$
Dichromate	$Cr_2O_7^{2-} + 6I^- + 14H^+ \rightleftharpoons 2\ Cr^{3+} + 3I_2 + 7H_2O$
Hydrogen peroxide	$H_2O_2 + 2H^+ + 2I^- \rightleftharpoons I_2 + 2H_2O$
Iodate	$IO_3^- + 5I^- + 6H^+ \rightleftharpoons 3I_2 + 3H_2O$
Nitrite	$2HNO_2 + 2I^- + 2H^+ \rightleftharpoons 2NO + I_2 + 2H_2O$
Oxygen	$O_2 + 4Mn(OH)_2 + 2H_2O \rightleftharpoons 4Mn(OH)_3$
	$2Mn(OH)_3 + 2I^- + 6H^+ \rightleftharpoons 2Mn^{2+} + I_2 + 6H_2O$
Ozone	$O_3 + 2I^- + 2H^+ \rightleftharpoons O_2 + I_2 + H_2O$
Periodate	$IO_4^- + 7I^- + 8H^+ \rightleftharpoons 4I_2 + 4H_2O$
Permanganate	$2MnO_4^- + 10I^- + 16H^+ \rightleftharpoons 2Mn^{2+} + 5I_2 + 8H_2O$

[14] B. Park, *Ind. Eng. Chem., Anal. Ed.*, **3**, 77 (1931).

[15] *Ibid.*

[16] W. R. Crowell, T. E. Hillis, S. P. Rittenberg, and R. F. Svenson, *Ind. Eng. Chem., Anal. Ed.*, **8**, 9 (1936).

[17] L. W. Winkler, *Ber.*, **21**, 2843 (1888); *Standard Methods for the Examination of Water and Sewage*, 9th ed., American Public Health Association, New York, 1946, p. 124.

oxygen dissolved in water. To the water sample is added an excess of a manganese(II) salt, sodium iodide, and sodium hydroxide. White $Mn(OH)_2$ is precipitated and is quickly oxidized to brown $Mn(OH)_3$:

$$4Mn(OH)_2(s) + O_2 + 2H_2O \longrightarrow 4Mn(OH)_3(s)$$

The solution is acidified and the $Mn(OH)_3$ oxidizes iodide to iodine,

$$2Mn(OH)_3(s) + 2I^- + 6H^+ \longrightarrow 2Mn^{2+} + I_2 + 6H_2O$$

The liberated iodine is titrated with a standard solution of sodium thiosulfate.

EXPERIMENT 4.15.   **Preparation and Standardization of a 0.1 *N* Sodium Thiosulfate Solution**

Procedure

Dissolve about 25 g of sodium thiosulfate pentahydrate crystals in 1 liter of water that has been recently boiled and cooled. Add about 0.2 g of sodium carbonate as a preservative and store in a clean bottle.

Standardization

(a) Potassium Dichromate. Weigh three portions of pure, dry potassium dichromate (Note 1) of about 0.2 g each into 500-ml Erlenmeyer flasks. Dissolve each sample in about 100 ml of water and add 4 ml of concentrated sulfuric acid. To the first sample carefully add 2 g of sodium carbonate (Note 2) with gentle swirling to liberate carbon dioxide. Then add 5 g of potassium iodide dissolved in about 5 ml of water (Note 3), swirl, cover the flask with a watch glass, and allow the solution to stand for 3 min (Note 4). Dilute the solution to about 200 ml and titrate with thiosulfate solution until the yellowish color of iodine has nearly disappeared (Note 5). Then add 5 ml of starch solution and continue the titration until 1 drop of titrant removes the blue color of the starch-iodine complex. The final solution will be clear emerald green, the color imparted by Cr(III) ion.

Treat the other samples in the same manner and calculate the normality of the thiosulfate solution. The average deviation should be about 1 to 3 parts per thousand.

Notes

1. The material can be obtained from the National Institute of Standards and Technology, but the best grade available from commercial supply houses is sufficiently pure for most purposes. Dry at 150°C if necessary.

2. The purpose of this is to remove air from the flask and lessen the danger of air oxidation of iodide. Do not add too much carbonate, as this will use up too much acid. The final concentration of acid is about 0.4 *M* if these quantities are used.

3. Do not allow this solution to stand as the iodide may be oxidized by air. The iodide should be free of iodate (see Note 3, Experiment 4.12) or a blank must be determined.

4. The reaction is somewhat slow but should be complete within this time.

5. Starch should not be added to a solution that contains a large quantity of iodine. The starch may be coagulated and the complex with iodine may not easily break up. A recurring end point will then be obtained.

(b) Copper. Secure some clean copper wire or foil, weigh three pieces of about 0.20 to 0.25 g each (Note 1), and place them in 250-ml Erlenmeyer flasks. Add to each flask 5 ml of 1:1 nitric acid and dissolve the copper by warming the solution on a steam bath or over a low flame. Add 25 ml of water and boil the solution for about 1 min. Then add 5 ml of urea solution (1 g in 20 ml of water) and continue boiling for another minute (Note 2). Cool the solution under the tap and neutralize the acid with 1:3 ammonia, adding the ammonia carefully until a pale blue precipitate of copper(II) hydroxide is obtained (Note 3). Now add 5 ml of glacial acetic acid and cool the solution if it is warm. To the first sample, add 3 g of potassium iodide, cover the flask with a watch glass, and allow to stand for 2 min. Then titrate with thiosulfate solution until the brownish color of iodine is almost gone (Note 4). Add 5 ml of starch solution and 2 g of potassium thiocyanate (Note 5). Swirl the flask for about 15 s and complete the titration, adding thiosulfate dropwise. At the end point the blue color of the solution disappears and the precipitate appears white, or slightly gray, when allowed to settle (Note 6).

Treat the second and third samples in the same manner and titrate with thiosulfate. Calculate the normality and copper titer of the thiosulfate.

Notes

1. The equivalent weight of copper is the atomic weight, 63.546.
2. Boiling removes oxides of nitrogen which result from the following reactions:

$$3Cu + 8HNO_3 \longrightarrow 3Cu(NO_3)_2 + 2NO + 4H_2O$$

$$2NO + O_2 \longrightarrow 2NO_2$$

Nitrous acid is formed by the reaction

$$NO_2 + NO + H_2O \longrightarrow 2HNO_2$$

and is decomposed by urea, according to the equation

$$2HNO_2 + (NH_2)_2CO \longrightarrow CO_2 + 2N_2 + 3H_2O$$

3. If excess ammonia is added, copper(II) hydroxide redissolves, forming the deep blue copper(II) ammonia complex, $Cu(NH_3)_4^{2+}$. If the excess of ammonia is large, a large quantity of ammonium acetate will be produced later; this may keep the reaction between copper(II) and iodide ions from being complete. The excess ammonia can be removed by boiling, and the precipitate will re-form.

4. After the first titration, calculate the approximate volume required for the other samples. Then titrate to within 0.5 ml of this volume before adding starch and thiocyanate.

5. Thiocyanate displaces adsorbed iodine from the precipitate.

6. The precipitate is seldom completely white. Do not continue addition of thiosulfate until the precipitate is white, since this will require considerable excess titrant.

EXPERIMENT 4.16.   **Determination of Copper in an Ore**

Procedure

Accurately weigh three samples of appropriate size (consult instructor) of the finely ground, dried ore into 250-ml beakers. Add 5 ml of concentrated hydrochloric acid and (slowly) 10 ml of concentrated nitric acid (Note 1). Cover the beaker with a watch glass and heat over a low flame until only a white residue of silica remains. Remove the watch glass and add 10 ml of 1:1 sulfuric acid. Then evaporate the solution (hood) until white fumes of sulfur trioxide appear (Note 2). Cool the solution and add carefully 25 ml of water.

Next add 5 ml of saturated bromine water and boil the solution gently for several minutes to expel excess bromine (Note 3). Then add 1:3 ammonium hydroxide carefully until a slight precipitate of iron(III) hydroxide is formed, or, if no iron is present, the deep blue color of the copper-ammonia complex is just formed. Now add 5 ml of glacial acetic acid, 2 g of ammonium acid fluoride (Notes 4 and 5), and stir until the iron(III) hydroxide redissolves.

Dissolve about 3 g of potassium iodide in 10 ml of water and add to the copper solution. Titrate the liberated iodine at once with thiosulfate until the brownish color of the iodine is almost gone (Note 6). Then add 5 ml of starch solution and 2 g of potassium thiocyanate (Note 7). Swirl the flask gently for about 15 sec and complete the titration, adding thiosulfate dropwise. At the end point the blue color of the solution disappears, and the precipitate appears white, or slightly gray, when allowed to settle (Note 8).

Treat the other two samples in the same manner and titrate with thiosulfate. Calculate the percentage of copper in the ore.

Notes

1. Some commercial unknowns are prepared from copper oxide and are easily dissolved by warming for 10 to 15 min with 15 ml of 1:2 sulfuric acid. No evaporation is then required. Continue the procedure with the addition of bromine water.

2. This removes nitric acid which would react with iodide ion.

3. Bromine is added to ensure that arsenic and antimony are in the +5 oxidation state. These elements may be reduced back to the +3 state by sulfur in a sulfide ore. Unless the excess bromine is removed, it will, of course, oxidize iodide ion. A test to see if bromine has been completely expelled is to hold in the vapors a piece of filter paper moistened with a starch solution which contains some potassium iodide. If the paper darkens, bromine is still present. If arsenic and antimony are absent, the bromine treatment can be omitted. Consult the instructor.

4. The container will be etched by hydrofluoric acid. It is preferable to transfer the solution at this point to a flask or beaker which is already etched or chipped. This container can be used for each sample after addition of flouride and then discarded.

5. Samples of copper oxide (Note 1) do not normally contain iron and it may not be

necessary to add fluoride. Add glacial acetic acid and continue the procedure. Consult the instructor.

6, 7, 8. See Notes 4, 5, and 6 of Experiment 4.15(b).

EXPERIMENT 4.17. Iodometric Determination of Hydrogen Peroxide

Procedure

Pipet 25 ml of the peroxide solution into a 250-ml volumetric flask, dilute to the mark, and mix thoroughly. Transfer a 25-ml aliquot of this solution to a 250-ml Erlenmeyer flask and add 8 ml of 1:6 sulfuric acid (about 6 N), 3 g of potassium iodide in 10 ml of water, and 3 drops of 3% ammonium molybdate solution. Titrate with thiosulfate until the brown color of iodine has almost disappeared. Then add 5 ml of starch solution and finish the titration to the disappearance of the deep blue color.

Titrate two other portions of the peroxide solution in the same manner. Calculate the weight of hydrogen peroxide in the original sample. Assuming that the weight of the sample was exactly 25 g, calculate the percentage of hydrogen peroxide by weight.

EXPERIMENT 4.18. Determination of Bleaching Power by Iodometry

Commercial bleaching products contain oxidizing agents such as hypochlorites or peroxides. The oxidizing power can be determined by the same procedure used in Experiment 4.17.

Procedure

Place an accurately measured sample of the bleach in a 250-ml Erlenmeyer flask. Liquid bleaches such as Clorox and Purex contain sodium hypochlorite, and 2.00 ml is a convenient sample if 0.1 N thiosulfate is used as the titrant. Solid products such as Clorox II and Snowy Bleach contain peroxides. A sample of 0.7 to 0.8 g is usually a convenient size for titration (Note 1). Add to the Erlenmeyer flask 75 ml of distilled water, 3 g of KI, 8 ml of 1:6 sulfuric acid, and 3 drops of 3% ammonium molybdate solution (Note 2). Titrate the liberated iodine with 0.1 N thiosulfate until the brown color of iodine has almost disappeared. Then add 5 ml of starch solution and finish the titration to the disappearance of the deep blue color.

Titrate at least two portions of each bleaching product being compared. The oxidizing ability of a bleach is usually reported as percent chlorine. That is, the calculation assumes that chlorine is the oxidizing agent, although in fact it may not be. Report the percent by weight of chlorine in each product, assuming that the liquid bleaches have a density of 1.000 g/ml. The equivalent weight of chlorine is the atomic weight.

Notes

1. The solids may not be homogeneous. If duplicate samples do give widely different results, a large sample can be taken, dissolved in a 500-ml volumetric flask, and 50-ml aliquots titrated.
2. Some of the bleaches react slowly with iodide ion. Molybdate ions catalyze the reaction. Alternatively, the solutions can be heated to about 60°C to speed up the reaction.

PERIODIC ACID

The compound paraperiodic acid, H_5IO_6, is a powerful oxidizing agent which is extremely useful in performing selective oxidations of organic compounds with certain functional groups. The standard potential of the reaction

$$H_5IO_6 + H^+ + 2e \rightleftharpoons IO_3^- + 3H_2O$$

is about $+1.6$ to 1.7 V.

Preparation and Standardization of Solutions

Three compounds are available commercially for the preparation of periodate solutions: H_5IO_6, paraperiodic acid; $NaIO_4$, sodium metaperiodate; and KIO_4, potassium metaperiodate. Of these compounds, $NaIO_4$, is generally preferred because of its relatively high solubility and ease of purification. Solutions as concentrated as 0.06 M can be prepared. Periodic acid solutions slowly oxidize water to oxygen and ozone. Solutions containing an excess of sulfuric acid are the most stable.

Periodate solutions are standardized by an iodometric procedure:

$$IO_4^- + 2I^- + H_2O \longrightarrow IO_3^- + I_2 + 2OH^-$$

Excess potassium iodide is added to an aliquot of the periodate and the liberated iodine is titrated with a standard solution of arsenic(III):

$$I_2 + AsO_2^- + 2H_2O \longrightarrow HAsO_4^{2-} + 2I^- + 3H^+$$

The solution is buffered at a pH of 8 to 9 with borax or sodium bicarbonate.

The Malaprade Reaction

In 1928 Malaprade[18] reported that periodic acid could be used for the selective oxidation of organic compounds with hydroxyl groups on *adjacent carbon atoms*. The reaction with ethylene glycol is

---

[18] L. Malaprade, *Compt. Rend.*, **186,** 382 (1928); *Bull. Soc. Chim. France*, (4)**43,** 683 (1928).

$$\begin{matrix} H_2C-OH \\ | \\ H_2C-OH \end{matrix} + H_4IO_6^- \longrightarrow 2H_2C{=}O + IO_3^- + 3H_2O$$

ethylene   formaldehyde
glycol

The carbon-carbon bond in the glycol is broken and two molecules of formaldehyde are produced. Excess periodate is used and after the reaction is complete, the excess can be determined by the iodometric procedure used for standardization. The reaction is carried out at room temperature, usually requiring from 30 min to 1 h. At higher temperatures undesired side reactions may occur and the selectivity of the periodate oxidation is not attained. Aqueous solutions which are neutral, slightly acid, or slightly basic can often be employed, although an organic solvent may be required if the compound is not soluble in water.

  Organic compounds which contain carbonyl groups ($>C{=}O$) on adjacent carbon atoms are also oxidized by periodate. The reaction with glyoxal is

$$\begin{matrix} H-C{=}O \\ | \\ H-C{=}O \end{matrix} + H_4IO_6^- \longrightarrow \begin{matrix} 2H-C{=}O \\ | \\ OH \end{matrix} + IO_3^- + H_2O$$

   glyoxal   formic acid

The carbon-carbon bond is broken and two molecules of formic acid are produced. A compound which contains a hydroxyl and a carbonyl group on adjacent carbon atoms is oxidized to an acid and an aldehyde:

$$\begin{matrix} H \\ | \\ CH_3-C-OH \\ | \\ CH_3-C{=}O \end{matrix} + H_4IO_6^- \longrightarrow \begin{matrix} H \\ | \\ CH_3-C{=}O \end{matrix} + \begin{matrix} CH_3-C{=}O \\ | \\ OH \end{matrix} + IO_3^- + 2H_2O$$

  acetoin   acetaldehyde acetic acid

  The compound glycerol is oxidized to 2 mol of formaldehyde and 1 mol of formic acid. The products can be rationalized by picturing the reaction as occurring in two steps. The first step produces two aldehydes,

$$\begin{matrix} H \\ | \\ H-C-OH \\ | \\ H-C-OH \\ | \\ H-C-OH \\ | \\ H \end{matrix} \longrightarrow H-C{=}O + \begin{matrix} H \\ | \\ H-C{=}O \\ | \\ H-C-OH \\ | \\ H \end{matrix} \longrightarrow H-C{=}O + \begin{matrix} H \\ | \\ H-C{=}O \\ | \\ OH \end{matrix}$$

glycerol   formaldehyde glycolic   formaldehyde formic
            aldehyde   acid

One of these aldehydes, glycolic aldehyde, contains a carbonyl and a hydroxyl group on adjacent carbon atoms and is therefore oxidized to an aldehyde and an acid.

The Malaprade reaction has found many applications in the determination of organic compounds. Glycerol and ethylene glycol can be easily determined by the reaction described above. Mixtures of glycerol, ethylene glycol and 1, 2-propyleneglycol have been determined in fats and waxes used in cosmetic materials. In addition, compounds containing a hydroxyl group and an amino group ($-NH_2$) on adjacent carbon atoms are readily oxidized to an aldehyde and ammonia. The ammonia can be distilled from the alkaline reaction solution and titrated with standard acid. This procedure has been used in the analysis of α-hydroxyamino compounds in various proteins.

EXPERIMENT 4.19. **Determination of Glycerol by Oxidation with Periodate[19]**

Directions are given below for the determination of glycerol by oxidation with periodate. In this reaction 1 mol of glycerol produces 1 mol of formic acid plus 2 mol of formaldehyde. The excess periodate is destroyed by allowing it to oxidize ethylene glycol to formaldehyde, and the formic acid is titrated with standard base using bromthymol blue indicator.

Procedure

Prepare a dilute solution of sulfuric acid by adding 30 ml of 0.2 N H_2SO_4 to about 300 ml of distilled water in a 600-ml beaker. Then dilute the solution to about 500 ml. Now dissolve 15 g of sodium metaperiodate, $NaIO_4$, in 250 ml of the dilute sulfuric acid solution. It should not be necessary to use heat to make the solid dissolve.

Prepare a 0.1% solution of bromthymol blue by dissolving 0.1 g of the indicator in 8 ml of 0.02 M NaOH and diluting to 100 ml with distilled water.

Weigh the sample containing glycerol (Note 1) and place it in a 500-ml Erlenmeyer flask. Add about 50 ml of distilled water and 6 drops of bromthymol blue to the flask. Prepare a blank in the same manner. Now carefully add 0.2 N H_2SO_4 to the sample and blank until the color of each solution is definitely green or greenish yellow. Then neutralize each solution with 0.05 N NaOH to a definite blue color free of green. Add with a pipet 50 ml of the periodic solution to each flask, cover with a watch glass, and allow the solutions to stand at room temperature for 1 h. Prepare a solution of ethylene glycol by mixing 30 ml of the glycol with 30 ml of water. Add to each the sample and the blank 10 ml of this solution and

[19] V. C. Mehlenbacher in *Organic Analysis,* Vol. 1, Interscience Publishers, Inc., New York, 1953.

allow the solutions to stand for 20 min. Dilute each solution to about 250 ml with distilled water and titrate with standard 0.1 N NaOH (Note 2). Repeat the determination on two additional samples (consult the instructor).

Subtract the volume of base used by the blank from that used by the sample. Then calculate the percentage of glycerol in the sample, noting that 1 mol of glycerol produces 1 mol of formic acid.

Notes

1. If the sample contains from 60 to 70% glycerol the amount weighed should be from 0.55 to 0.75 g; if 50 to 60%, from 0.65 to 0.85 g; if 40 to 50%, from 0.80 to 1.00 g; if 30 to 40%, from 0.90 to 1.30 g. See the reference in footnote 19 for a complete table of recommended sample sizes.

2. If a pH meter is available, it is recommended[20] that the sample be titrated to a pH of 8.1 and the blank to a pH of 6.5.

---

[20] *Ibid.*

5

Precipitation
and Complex
Ion Titrations

The theory of titrations involving the formation of complexes and precipitates is discussed in Chapters 8 and 9 of the text. In this chapter laboratory directions are given for precipitation titrations involving the reaction of silver cation with chloride and thiocyanate anions. The Mohr, Volhard, and Fajans methods are illustrated. Titrations involving complex formation include the 'reaction of silver ion with cyanide, mercury(II) ion with chloride, and the use of EDTA in determining the hardness of water and calcium in a drug.

EXPERIMENT 5.1.   **Preparation of 0.1 M Solutions of Silver Nitrate and
                       Potassium Thiocyanate**

Silver nitrate can be weighed as a primary standard. However, a solution of approximately the desired molarity is usually prepared and then standardized using the indicator which will be employed in the analysis.

Procedure

(a) *Silver Nitrate.* Weigh on a trip balance about 8.5 g of silver nitrate (Note 1), dissolve the salt in distilled water in a beaker, and transfer the solution to a clean bottle (Note 2). Dilute the solution to 500 ml with distilled water, mix thoroughly, and label the bottle. Protect the solution from the sunlight as much as possible.

(b) Potassium Thiocyanate. If the Volhard method is to be used, prepare a 0.1 M thiocyanate solution by weighing about 4.9 g of potassium thiocyanate (Note 3) and dissolving the salt in water. Dilute the solution to about 500 ml and store in a clean bottle.

Notes

1. The molecular weight is 169.87. To prepare 500 ml of a 0.1 M solution, 8.5 g is needed which is ample for standardization and determination of an unknown. If 1 liter of solution is desired, take 17 g.
2. A brown-glass bottle is recommended to protect the solution from sunlight.
3. The molecular weight is 97.18. If 1 liter of the solution is desired, take 9.7 g of the salt.

EXPERIMENT 5.2. Standardization of Silver Nitrate and Potassium Thiocyanate Solutions

Sodium or potassium chloride can be used as a primary standard for a silver nitrate solution. Either potassium chromate (Mohr method) or dichlorofluorescein (Fajans, or adsorption indicator, method) can be used as the indicator. Directions are given for both indicators below.

After standardization of the silver nitrate, this solution can be used to standardize the potassium thiocyanate solution. The two solutions are titrated directly, using iron(III) ion as the indicator.

Procedure

(a) Mohr Method. Weigh accurately three samples of pure, dry sodium chloride of about 0.20 to 0.25 g each (Notes 1 and 2) into 250-ml Erlenmeyer (Note 3) flasks. Dissolve each sample in 50 ml of distilled water and add 2 ml of 0.1 M potassium chromate solution (Note 4). Titrate the first sample with silver nitrate, swirling the solution constantly, until the reddish color of silver chromate begins to spread more widely through the solution, showing that the end point is almost reached. The formation of clumps of silver chloride is also an indicator that the end point is near. Continue the addition of silver nitrate dropwise until there is a permanent color change from the yellow of the chromate ion to the reddish color of silver chromate precipitate. Run an indicator blank (Note 5) if desired (consult the instructor).

Titrate the other two samples in the same manner. Calculate the molarity of the silver nitrate solution and its chloride titer.

(b) Fajans Method. Weigh accurately three samples of pure, dry sodium

chloride of about 0.20 to 0.25 g each (Notes 1 and 2) into 250-ml Erlenmeyer flasks. Dissolve each sample in 50 ml of water and add 10 drops to dichlorofluorescein indicator (Note 6) and 0.1 g dextrin (Note 7). Titrate the first sample with silver nitrate to the point where the color of the dispersed silver chloride changes from yellowish white to a definite pink (Note 8). The color change is reversible, and back-titration can be done with a standard sodium chloride solution. Titrate the other two samples in the same manner. Calculate the molarity of the silver nitrate solution and its chloride titer.

Notes

1. The molecular weight is 58.44. If potassium chloride, molecular weight 74.55, is used, take 0.25 to 0.30 g. The method of aliquot portions may be used if desired. Weigh accurately 1.1 to 1.2 g of sodium chloride (1.5 to 1.6 g of potassium chloride), dissolve the salt in a 250-ml volumetric flask, and withdraw 50-ml aliquots for titration.
2. Dry the salt at 120°C for at least 2 h. For complete removal of the last traces of water, these salts should be heated to about 500 to 600°C in an electric furnace. This is not necessary except for very precise work.
3. Porcelain casseroles are sometimes recommended. The reddish color of silver chromate is more readily distinguished against a white background.
4. Dissolve 19.4 g in 1 liter of water.
5. To determine the indicator blank, add 2 ml of the indicator to about 100 ml of water to which is added a few tenths of a gram of chloride-free calcium carbonate. This gives a turbidity similar to that in the actual titration. Swirl the solution and add silver nitrate dropwise until the color matches that of the solution that was titrated. The blank should not be larger than about 0.05 ml.
6. This solution is prepared by dissolving 0.1 g of sodium dichlorofluoresceinate in 100 ml of water (or 0.1 g of dichlorofluorescein in 100 ml of 70% alcohol). Fluorescein, or its sodium salt, can be used in place of dichlorofluorescein.
7. The function of dextrin is to prevent coagulation of colloidal silver chloride.
8. The end point is easier to detect in diffuse light. Avoid direct sunlight.

(c) *Titration of Silver Nitrate with Potassium Thiocyanate.* Pipet 25 ml of standard silver nitrate solution into a 250-ml Erlenmeyer flask. Add 5 ml of 1:1 nitric acid (Note 1) and 1 ml of iron(III) alum solution as the indicator (Note 2). Titrate with thiocyanate, swirling the solution constantly, until the reddish-brown color begins to spread throughout the solution. Then add the thiocyanate dropwise, shaking the solution thoroughly between addition of drops. The end point is marked by the permanent appearance of the reddish color of the iron-thiocyanate complex.

Titrate two additional portions of the silver nitrate solution with thiocyanate. Calculate the molarity of the thiocyanate solution.

Notes

1. If the nitric acid has a yellow tinge indicating the presence of oxides of nitrogen, boil the solution until the oxides are expelled.
2. This is a saturated solution of iron(III) ammonium sulfate in 1 M nitric acid.

EXPERIMENT 5.3. **Determination of Chloride**

Chloride can be determined by titration with standard silver nitrate. A direct titration using either chromate ion or dichlorofluorescein as the indicator can be employed. Alternatively, excess standard silver solution can be added to the unknown and the excess titrated with standard potassium thiocyanate. Directions are given below for all three procedures. It is assumed that the sample is water soluble.

Procedure

(a) Mohr Method. Accurately weigh three samples of the dried material (Note 1) into three 250-ml Erlenmeyer flasks (Note 2). Dissolve each sample in about 50 ml of distilled water (Note 3) and add 2 ml of 0.1 M potassium chromate solution. Titrate the first sample with standard silver nitrate as directed in Experiment 5.2(a).

Titrate the other two samples in the same manner. Calculate the percentage of chloride in the sample.

(b) Fajans Method. Accurately weigh three samples of the dried material of appropriate size (Note 1) into three 250-ml Erlenmeyer flasks. Dissolve each sample in about 50 ml of distilled water (Note 3) and add 10 drops of dichlorofluorescein indicator and 0.1 g of dextrin. Titrate the first sample with standard silver nitrate as directed in Experiment 5.2(b).

Titrate the other two samples in the same manner. Calculate the percentage of chloride in the sample.

Notes

1. Consult the instructor regarding the size of sample. A large sample may be dissolved in a 250-ml volumetric flask and aliquot portions titrated, if desired.
2. Porcelain casseroles are sometimes recommended.
3. The solution should be nearly neutral. Dissolve a small portion of the material in about 10 ml of water and place a drop of the solution on a piece of litmus paper. If the solution is basic, add 1 drop phenolphthalein to each solution and then add dropwise dilute nitric acid (about 1 drop to 150 ml of water) until the pink color of the indicator is just discharged. If the solution is acidic, add 1 drop of phenolphthalein to each solution and then add dropwise 0.1 N sodium hydroxide until the solution is barely pink. Then add 1 or 2 drops of dilute nitric acid until the pink color of the indicator is just discharged.

(c) Volhard Method. Accurately weigh three samples of appropriate size (Note 1) into three 250-ml Erlenmeyer flasks. Dissolve each sample in about 50 ml of distilled water and then add 5 ml of 1:1 nitric acid (Note 2). To the first sample add standard silver nitrate solution from a buret until an excess of about 5 ml is present (Note 3). Add 1 to 2 ml of nitrobenzene (Note 4), stopper the flask with a rubber stopper, and shake the flask vigorously until the silver chloride is

well coagulated (about 30 s). Now add 1 ml of iron(III) alum indicator (Note 5) and titrate the excess silver nitrate with standard potassium thiocyanate solution. The end point is marked by the permanent appearance of the reddish color of the iron-thiocyanate complex.

Titrate the other two samples in the same manner. Calculate the percentage of chloride in the sample.

Notes

1. Consult the instructor. If desired, a larger sample may be taken, dissolved in a 250-ml volumetric flask, and aliquot portions titrated.

2. See Note 1 of Experiment 5.2(c).

3. The silver chloride coagulates near the equivalence point. Shake the solution well, allow it to stand for a few moments for the precipitate to settle, and then add a few drops of silver nitrate to the supernatant liquid. If no precipitate forms, silver nitrate is in excess.

4. Instead of adding nitrobenzene, one may filter off the silver chloride with a Gooch or sintered glass crucible. The precipitate is then washed with 1% nitric acid solution, the washings being added to the filtrate. The indicator is then added to the filtrate and the latter titrated with potassium thiocyanate.

5. See Note 2 of Experiment 5.2(c).

EXPERIMENT 5.4.   **Determination of Silver in an Alloy**

Silver can be determined by direct titration with potassium thiocyanate using iron(III) ion as the indicator.

Procedure

Accurately weigh three samples of a silver alloy of appropriate size (consult the instructor) and place each in a 250-ml Erlenmeyer flask. Dissolve the first sample in 15 ml of 1 : 1 nitric acid. Boil the solution until all the oxides of nitrogen are removed and then dilute the solution to about 50 ml. Add 1 ml of iron(III) alum indicator (Note) and titrate with standard thiocyanate solution to the appearance of the reddish color of the iron-thiocyanate complex.

Dissolve and titrate the other two samples in the same manner. Calculate the percentage of silver in the alloy.

Note.

See Note 2 of Experiment 5.2(c).

EXPERIMENT 5.5.   **Determination of Cyanide by the Liebig Method**

The standard silver nitrate solution can be used to determine cyanide according to the Leibig method. There is no interference from chloride, bromide, or iodide ions since the silver salts of these anions are soluble in excess cyanide. Iodide ion is used as the indicator in the procedure given below.

Procedure

Secure a sample from the instructor (Note 1). Weigh accurately three 125-ml Erlenmeyer flasks that are clean, dry, and fitted with cork stoppers. Then add to each flask about 8 or 10 ml of the cyanide solution (Note 2) and reweigh the flask and contents. Add 3 ml of concentrated ammonia and about 0.1 g of potassium iodide to each flask and dilute to about 50 ml with distilled water. Titrate the first sample with silver nitrate, swirling the flask constantly, until 1 drop of titrant produces a faint permanent opalescence in the solution (Note 3).

Titrate the other two samples in the same manner. Calculate the percentage by weight of cyanide in the solution.

Notes

1. Since sodium and potassium cyanides are difficult to dry, the unknown sample may be dispensed as a solution. The solution should be titrated relatively soon since it slowly decomposes. *Caution:* Do not pipet any of this solution by mouth because it is extremely poisonous. Also, do not acidify the solution under any circumstances.

2. This volume should contain 7 to 8 mmol of cyanide. Remember that each millimole of silver reacts with 2 mmol of cyanide.

3. After finishing the titration, add ammonia to the flask and pour the solution down a drain. Rinse the flask thoroughly with a large quantity of water.

EXPERIMENT 5.6.   **Preparation and Standardization of Sodium-EDTA Solution**

Titrations involving the use of the chelating agent EDTA are described in Chapter 8 of the text. Directions are given below for the preparation of a 0.01 *M* solution of sodium–EDTA and the standardization against calcium chloride.

Procedure

Weigh about 4 g of disodium dihydrogen EDTA dihydrate and about 0.1 g of $MgCl_2 \cdot 6H_2O$ into a clean 400-ml beaker. Dissolve the solids in water, transfer the solution to a clean 1-liter bottle, and dilute to about 1 liter (Note 1). Mix the solution thoroughly and label the bottle.

Prepare a standard calcium chloride solution as follows. Weigh accurately about 0.4 g of primary standard calcium carbonate that has been previously dried at 100°C. Transfer the solid to a 500-ml volumetric flask, using about 100 ml of water. Add 1:1 hydrochloric acid dropwise until effervescence ceases and the solution is clear. Dilute with water to the mark and mix the solution thoroughly.

Pipet a 50-ml portion of the calcium chloride solution into a 250-ml Erlenmeyer flask and add 5 ml of an ammonia–ammonium chloride buffer solution (Note 2). Then add 5 drops of Eriochrome Black T indicator (Note 3). Titrate carefully with the EDTA solution to the point where the color changes from wine-red to pure blue. No tinge of red should remain in the solution.

Repeat the titration with two other aliquots of the calcium solution. Calculate the molarity of the EDTA solution and the calcium carbonate titer.

Notes

1. If the solution is turbid, add a few drops of 0.1 M sodium hydroxide solution until the solution is clear.

2. Prepare this solution by dissolving about 6.75 g of ammonium chloride in 57 ml of concentrated ammonia and diluting to 100 ml. The pH of the buffer is slightly above 10.

3. Prepare by dissolving about 0.5 g of reagent-grade Eriochrome Black T in 100 ml of alcohol. If the solution is to be kept, date the bottle. It is recommended that solutions older than 6 weeks to 2 months not be used. Alternatively, the indicator may be used as a solid, which has a much longer shelf life. It is prepared by grinding 100 mg of the indicator into a mixture of 10 g of NaCl and 10 g of hydroxylamine hydrochloride. A small amount of the solid mixture is added to each titration flask with a spatula.

Alternatively, Calmagite may be used as the indicator. A solution is prepared by dissolving 0.05 g of the indicator in 100 ml of water. Add 4 drops of the indicator to each flask. The color change is from red to blue, as with Eriochrome Black T.

EXPERIMENT 5.7.   Determination of the Total Hardness of Water

The standard EDTA solution can be used to determine the total hardness of water.

Procedure

Obtain the water to be analyzed from the instructor and pipet a portion into each of three 250-ml Erlenmeyer flasks (Note 1). To the first sample add 1.0 ml of the buffer solution (Note 2) and 5 drops of indicator solution (Note 3). Titrate with the standard EDTA solution to a color change of wine red to pure blue.

Repeat the procedure on the other two portions of water. Calculate the total hardness of the water as parts per million (ppm) of calcium carbonate. This is done as follows:

$$\text{Volume EDTA (ml)} \times \text{CaCO}_3 \text{ titer (mg/ml)} = \text{mg CaCO}_3$$

$$\frac{1000 \text{ ml/liter} \times \text{mg CaCO}_3}{\text{ml sample}} = \text{mg CaCO}_3/\text{liter, or ppm}$$

Notes

1. The volume of water titrated should be chosen so that 40 to 50 ml of the EDTA solution will be used for titration.
2. See Note 2 of Experiment 5.6.
3. See Note 3 of Experiment 5.6.

EXPERIMENT 5.8. Determination of Calcium in Calcium Gluconate[1]

Of all the minerals in the body, calcium is the most abundant. Ninety-eight percent of the body's calcium is in the bones and 1% is in the teeth. The other 1% which is present in all the other tissues is essential for certain metabolic reactions such as the contraction of muscles.

Blood serum contains 4.5 to 5.5 mmol/liter of calcium. A significant increase or decrease produces pathological symptoms. Hypocalcemia results from a deficiency of parathyroid hormone, resistance to the hormone, or deficiency of vitamin D. Severe tetany occurs when the calcium concentration falls to 3.5 mmol/liter.

Standard treatment of a hypocalcemic patient consists of intravenous injection of calcium gluconate, along with parathyroid extract and vitamin D. In less acute cases 1 to 5 g of calcium gluconate may be given orally three times daily.

Calcium gluconate is a pleasant-tasting white compound, available as tablets or in solution. The anion is that of a weak organic acid and is not as likely to evoke acidosis as are ions such as chloride. Calcium gluconate thus is accepted as the prototype for calcium preparations in medicine.

Procedure

Calcium in calcium gluconate will be determined by titration with a standard solution of EDTA. It is recommended that the concentration of EDTA be about 0.025 M. Follow the procedure for preparing the solution given in Experiment 5.6, except use about 10 g of $Na_2H_2Y \cdot 2H_2O$. If some solid fails to dissolve, add NaOH pellets one at a time and shake the solution until no undissolved solid remains.

Prepare the standard calcium chloride solution as in Experiment 5.6, except use about 1.0 g of primary standard-grade calcium carbonate. Titrate three 50-ml aliquot portions of this solution with the EDTA solution as directed in Experiment 5.6.

Secure your calcium gluconate tablets from the instructor (Note 1). Crush them with the blade of a stainless steel spatula and mix the powder. Dry the powder at 100°C for 1 h. Weigh accurately a 0.4-g sample into each of three clean 250-ml Erlenmeyer flasks and dissolve each sample in 50 ml of distilled water. Some

[1] This experiment was designed by Hubert L. Youmans of Western Carolina University, who has kindly given us permission to use it here.

heating may be required. To the first sample add 5 ml of the ammonia–ammonium chloride buffer (Note 2, Experiment 5.6) and 5 drops of Eriochrome Black T indicator (Note 2). Titrate with the standard EDTA solution to a color change of wine red to pure blue.

Repeat the procedure on the other two samples. Calculate the percentage of calcium in the drug.

Notes

1. The tablets are available without a prescription from drugstores.
2. See Note 3 of Experiment 5.6.

EXPERIMENT 5.9.   Determination of Chloride by Titration with Mercury(II) Nitrate

The fact that chloride ion can be determined by titration with a standard solution of mercury(II) ion has been known for many years. However, the method has not been commonly used in the introductory laboratory because of the convenience of the method involving the precipitation of silver chloride. In recent years, as silver nitrate has become increasingly expensive, the mercurimetric method has gained in popularity.

The procedure given here is based on the work of Roberts,[2] who studied diphenyl carbazide as the indicator for the titration. Diphenyl carbazide is an acid-base indicator (HIn) which changes from a light yellow color in acid solution to deep orange in alkaline solution. Apparently the alkaline form of the indicator (In$^-$) reacts with Hg^{2+} to form a deep blue-violet complex. The concentration of In$^-$ must be kept sufficiently low by control of pH so that the reaction with Hg^{2+} does not occur too early in the titration. Satisfactory results are obtained if the pH is between 1.5 and 2.0. Directions are given here for the preparation and standardization of 500 ml of a 0.05 M (0.10 N) solution of mercury(II) nitrate and for the determination of chloride in a soluble salt. The proper pH is obtained by titrating in 0.02 M HNO$_3$.

Procedure

(a) *Preparation and Standardization of Mercury(II) Nitrate.* Weigh about 8.6 g of Hg (NO$_3$)$_2$ · H$_2$O (or 8.1 g of the anhydrous salt) into a clean 400-ml beaker. Dissolve the solid in 0.02 M HNO$_3$ (Note 1), transfer the solution to a clean bottle, and dilute the solution to 500 ml with the nitric acid solution.

Weigh accurately three samples of pure, dry sodium chloride of about 0.20 to 0.25 g each (Note 2) into 250-ml Erlenmeyer flasks. Dissolve each sample in 80 ml of 0.02 M HNO$_3$ and add 5 drops of diphenyl carbazide indicator solution

[2] I. Roberts, *Ind. Eng. Chem.*, *Anal. Ed.*, **8**, 365 (1935).

(Note 3). Titrate the first sample with $Hg(NO_3)_2$ solution, stopping the addition of titrant at the first appearance of a permanent pink color (Note 4).

Titrate the other two samples in the same manner. Calculate the molarity of the $Hg(NO_3)_2$ solution and its chloride titer.

(b) Determination of Chloride. Accurately weigh three samples of the dried unknown (Note 5) into three 250-ml Erlenmeyer flasks. Dissolve each sample in 80 ml of 0.02 M HNO_3 and add 5 drops of diphenyl carbazide indicator. Titrate the first sample with standard $Hg(NO_3)_2$ solution to the first appearance of a permanent pink color (Note 4).

Titrate the other two samples in the same manner. Calculate the percentage of chloride in the sample.

Notes

1. Add about 1.5 ml of concentrated HNO_3 to 1 liter of distilled water and dilute to about 1200 ml. This should be sufficient acid for the titration of three standards and three unknowns.

2. Dry the salt at 120°C for at least 2 h.

3. Saturate 50 ml of 95% ethyl alcohol with diphenyl carbazide. The solution gradually turns red on standing, but this change causes no trouble.[3]

4. Roberts[3] reported that the pink color develops about 5 drops before the formation of the deep blue-violet color. The latter color change is from a light violet to the deep violet, and some chemists find it easier to judge the first color change than the second. Either color change can be taken as the end point if it is used both for the standardization and the unknown.

5. Consult the instructor regarding the size of sample. A large sample may be dissolved in a 250-ml volumetric flask and aliquot portions titrated, if desired.

---

[3] *Ibid.*, p. 365.

6

Some Simple Gravimetric Determinations

The principles of separation by precipitation are discussed in Chapters 4 and 9 of the text. In this chapter we give directions for the precipitation of silver chloride, barium sulfate, and hydrous iron(III) oxide. These three compounds illustrate the common types of precipitates: curdy, crystalline, and gelatinous. Some discussion of the properties of the precipitates and the errors involved in their use is also included. Directions are also given for the use of an organic precipitant, dimethylglyoxime.

SILVER CHLORIDE

The precipitation of silver chloride can be used for the gravimetric determination of chloride and silver as well as the determination of chlorine in various states of oxidation. Since it is one of the simplest precipitations to carry out and since it gives excellent results even for beginners, it is usually the first example given the student in the gravimetric laboratory.

Properties

The solubility of silver chloride in water is slight (about 1.4 mg/liter at 20°C and 22 mg/liter at 100°C). The solubility is further decreased by the addition of excess precipitant and hence losses due to solubility in the mother liquor are negligibly small.

Silver chloride precipitates in curds or lumps resulting from the coagulation of colloidal material. After addition of excess precipitant, the solution is normally heated and stirred well in order to coagulate the colloid completely. The curdy precipitate has little tendency to occlude foreign substances and is hence obtained in a fairly pure condition. Nitric acid is added to the solution to prevent precipitation of other silver salts such as the carbonate or phosphate.

The precipitate is easily filtered and washed free of the mother liquor. A little nitric acid is added to the wash water in order to prevent peptization of the precipitate. The acid is volatilized when the precipitate is dried. The precipitate is not heated to high temperatures because the salt melts at 455°C and may be lost by volatilization. Silver chloride is easily reduced by carbon of filter paper and hence is normally filtered through a Gooch, sintered glass, or porous porcelain crucible. It is then dried at 110 to 130°C. Since the solid does not hold water very strongly, this temperature is sufficient for drying except for very precise determinations such as atomic weights.

Errors

As previously mentioned, the precipitate of silver chloride gives excellent analytical results. Nevertheless, it is important to be aware of certain errors that may decrease the accuracy of the determination.

As mentioned above, solubility losses are negligible. However, alkali and ammonium salts, as well as large concentrations of acids, should be avoided since they increase the solubility.

Direct or diffuse sunlight darkens a silver chloride precipitate by decomposing the material into free silver and chlorine:

$$2AgCl \longrightarrow 2Ag + Cl_2(g)$$

The extent of this reaction is negligible unless the precipitate is exposed to direct sunlight and stirred occasionally. Lundell and Hoffman[1] found that in the determination of silver, with excess chloride added, this decomposition led to low results. In the determination of chloride, however, with excess silver ion present, additional silver chloride is precipitated by chloride ion formed from the interaction of chlorine and water, and the weight of precipitate is made high. After filtration, of course, this decomposition would lead to low results because the chlorine simply escapes.

In any case, the amount of decomposition of silver chloride with the usual laboratory lighting is slight and can be kept small by shielding the precipitate as much as possible from strong light.

Other Applications

In addition to the determination of silver and chloride, the precipitation of silver chloride can be used to determine chlorine in oxidation states other than −1.

[1]G. E. F. Lundell and J. I. Hoffman, *J. Res. Natl. Bur. Standards*, **4**, 109 (1930).

Hypochlorites, chlorites, and chlorates may be determined by first reducing these ions to chloride and then precipitating silver chloride. Sulfurous acid or sodium nitrite can be used as the reducing agent for all these substances. Chlorine in organic compounds is often determined by this precipitation after the organic chlorine is converted into sodium chloride by fusion with sodium peroxide.

Bromide and iodide may be determined by precipitation of their silver salts. Also, oxygen-containing anions such as hypobromite, bromate, hypoiodite, iodate, and periodate can first be reduced to bromide or iodide and then precipitated as the silver salts.

EXPERIMENT 6.1. Determination of Chlorine in a Soluble Chloride

The usual samples given to students are readily soluble in water, and no interfering ions are present. Hence this is a very simple determination that requires a relatively short time, and it is quite suitable for the introduction to gravimetric technique.

Procedure

The sample should be dried for 1 to 2 h in an oven at 100 to 120°C. Clean three (Note 1) porous-bottom filtering crucibles (either sintered glass or porous porcelain) of "medium" porosity. Much of the dirt may be removed with detergents, although acids or other "strong" reagents may be necessary at times, depending upon the nature of the contamination. It is often well to draw a little concentrated nitric acid slowly through the porous filtering disk to eliminate certain material with which the disk may be impregnated. It is not wise to use "cleaning solution" on a porous-bottom crucible. The cleaned and thoroughly rinsed crucibles should be dried to constant weight at the same temperature at which the silver chloride is later to be dried. An oven is recommended, and temperatures of 100 to 150°C are suitable for ordinary student work.

Weigh out accurately three portions of the dried sample of about 0.5 to 0.7 g each (Note 2). Dissolve each portion in a 250-ml beaker, using 100 to 150 ml of water to which about 1 ml of concentrated nitric acid has been added. Prepare an approximately 0.1 M solution of silver nitrate (15 to 20 mg of $AgNO_3$/ml). Now heat the first of the chloride solutions to boiling, and with constant stirring add silver nitrate slowly to precipitate silver chloride. Obviously, in the case of an unknown chloride solution, the quantity of silver nitrate to be added cannot be specified. The student himself must determine when precipitation is complete. This is done by adding the silver nitrate in small portions, stirring vigorously, allowing the precipitate to settle somewhat, and noting whether a new cloud of precipitate appears upon further addition. After precipitation is complete, add about 10% more silver nitrate solution.

After the precipitate has coagulated well, remove the beaker from the heat, cover it with a watch glass, and set it aside to cool in the laboratory bench (for protection from light) for at least 1 h. The other two samples are precipitated in the same way. After the first one has been done, the student will know the approximate amount of silver nitrate solution to add to the others and hence can proceed more rapidly.

After the solution has cooled, it is filtered through a weighed crucible with suction, retaining the bulk of the precipitate in the beaker. It is wise to test the filtrate once again for completeness of precipitation, using a few drops of silver nitrate solution. The precipitate in the beaker is then washed by decantation with three 25-ml portions of 0.01 M nitric acid (about 2 drops of concentrated nitric acid in 100 ml of water). The washings, of course, are poured through the filter. Finally, the precipitate is stirred up in a small portion of wash solution and transferred into the crucible. Any precipitate remaining in the beaker is carefully rinsed out into the crucible, using a rubber policeman if necessary to remove precipitate adhering to the walls of the beaker. Now wash the precipitate in the crucible three or four times more with small portions of wash solution, allowing it to drain each time. Collect the last portion of wash solution separately and test it for the absence of silver ion with a drop of hydrochloric acid. Finally, drain the crucible completely with strong suction and place it in a covered beaker for drying in the oven. The watch glass covering the beaker should be raised with small glass hooks so as to permit the circulation of air over the precipitate.

After the samples have dried for about 2 h, cool them in a desiccator and weigh. Return them to the oven for about 30 min, cool them in a desiccator, and reweigh. Drying may be considered complete if no more than 0.4 mg was lost during the second drying period.

Calculate the percentage of chlorine in the unknown sample. The gravimetric factor is 0.2474. Report the results in the manner prescribed by the instructor.

Notes

1. Consult the instructor concerning the number of replicates to be run. Ordinarily it is considered adequate to perform the analysis in triplicate.

2. The sample size should be such that a convenient quantity of precipitate is obtained. Consult the instructor. Portions of 0.5 to 0.7 g are appropriate for most student unknowns purchased currently.

BARIUM SULFATE

The precipitation of barium sulfate can be used for the gravimetric determination of barium and sulfate, as well as for the determination of sulfur in various states of oxidation. The precipitation is simple but it is difficult to obtain accurate results because the precipitate is usually impure. Nevertheless, the determination is commonly used as a student exercise.

Properties

The solubility of barium sulfate in water is small, about 3 mg/liter at 26°C. The solubility is less, of course, in the presence of excess precipitating agent, and losses due to solubility are small. Although the solubility of barium sulfate is greater in acid than in water, the precipitation is carried out in about 0.01 N hydrochloric acid for the following reasons: (1) Larger particles of precipitate are obtained. (2) A purer precipitate results. (3) The precipitation of such salts as $BaCO_3$ is prevented. The increased solubility in acid solution results from the formation of the bisulfate ion, HSO_4^- (dissociation constant about 0.01). Despite this increased solubility, the precipitation is still quantitative if the concentration of acid is as small as 0.01 M.

Barium sulfate precipitates as a crystalline solid and conditions must be adjusted in order to obtain particles as large as possible and to minimize coprecipitation. The precipitation is carried out by rapid mixing of dilute solutions under conditions of increased solubility. The solubility is increased by raising the temperature and by using an acidic solution (see above). The precipitate is then digested at an elevated temperature for one to 2 h. When the precipitation is carried out in this manner, the particles of precipitate are larger and purer than otherwise would be obtained. Nevertheless, it is still necessary to use a "slow" filter paper such as Whatman No. 42.

Errors

As previously pointed out, coprecipitation of foreign substances with barium sulfate is very pronounced. The anions most strongly coprecipitated are nitrate and chlorate. Cations, particularly divalent and trivalent ones that form slightly soluble sulfates, are strongly coprecipitated, iron(III) ion being one of the most prominent examples. The procedures summarized in Chapter 4 of the text are employed where practical to minimize coprecipitation. The digestion process employed to increase particle size also leads to some purification. Reprecipitation is not employed since a suitable solvent is not available.

The impurities that are coprecipitated can lead to either high or low results. In the determination of sulfate, coprecipitation of barium chloride leads to high results since the material should not come down at all. Coprecipitation of sulfuric acid or foreign metal sulfates causes low results; sulfuric acid is volatilized during ignition of the precipitate, and most metals are lighter than barium. In addition, some metal sulfates, such as iron(III) sulfate, are changed into oxides upon ignition. In the determination of barium, the coprecipitation of barium chloride causes low results since chloride is lighter than sulfate. All foreign sulfates cause high results because these substances should not be precipitated at all.

Barium sulfate is normally filtered with filter paper and washed with hot water. The filter paper must be burned off carefully with a plentiful supply of air.

The sulfate is reduced rather readily by the carbon of the paper:

$$BaSO_4 + 4C \longrightarrow 4CO(g) + BaS$$

If this reduction occurs, the results are low and usually the precision is poor. The precipitate can be reconverted to the sulfate by moistening it with sulfuric acid and reigniting. A porous porcelain crucible can be employed instead of paper.

Other Applications

The sulfur in sulfides, sulfites, thiosulfates, and tetrathionates can be determined by oxidizing the sulfur to sulfate and then precipitating barium sulfate. Permanganate is often used to effect the oxidation. Sulfur in organic compounds is determined by oxidizing the element to sulfate with sodium peroxide. Ores of sulfur, such as pyrites (FeS_2) and chalcopyrite ($CuFeS_2$), may be fused with sodium peroxide to oxidize the sulfur to sulfate.

The other cations often precipitated as sulfates are lead and strontium. Both of these sulfates are more soluble than barium sulfate. Alcohol is added in the determination of strontium to decrease the solubility of the sulfate. The determination of lead in brass is carried out by precipitation of lead sulfate.

EXPERIMENT 6.2.   Determination of Sulfur in a Soluble Sulfate

Procedure

The sample should be dried in an oven at 100 to 120°C (Note 1). Weigh out accurately three portions of about 0.5 to 0.8 g each (Note 2) and dissolve each sample in about 200 ml of distilled water and 1 ml of concentrated hydrochloric acid. Prepare a 5% solution of barium chloride by dissolving 5 g of $BaCl_2 \cdot 2H_2O$ in 100 ml of water. Calculate the volume of this solution which will be required to precipitate the sulfate in each sample, including a 10% excess (Note 2). Measure this volume of solution into a clean beaker using a graduated cylinder.

Now heat both the sample solution and the solution of barium chloride nearly to boiling. Pour the hot barium chloride solution quickly but carefully into the hot sample solution and stir vigorously. Allow the precipitate to settle and test the supernatant liquid for completeness of precipitation by adding a few more drops of the barium chloride solution.

After precipitation is complete, cover the beaker with a watch glass and allow the precipitate to digest for 1 to 2 h, keeping the solution hot (80 to 90°C) on a steam bath, hot plate, or using a low flame.

Either a fine porous porcelain filtering crucible or a funnel with filter paper can be used to collect the precipitate. The ignition temperature is too high to permit the use of sintered glass crucibles. If filter paper is used, the slow type such as

Whatman No. 42 should be selected unless the precipitate has been unusually well digested, in which case No. 40 may suffice.

Three crucibles, porous porcelain (or, if filter paper is to be used, ordinary porcelain), and lids are cleaned, rinsed, and heated to constant weight. Use the highest temperature of the Tirrill burner for an ordinary crucible. Remember that a porous porcelain crucible must be heated within another crucible to prevent damage and the admission of gases from the flame through the porous bottom. Thus the full temperature of a Meker burner is required in this case.

The directions for washing and filtering the precipitate apply whichever type of filter is used. The solution is to be hot at the time of filtration. Decant the clear supernatant solution through the filter. Discard the clear filtrate so that if the precipitate later runs through the filter only a small volume need be refiltered. Then rinse the precipitate into the funnel or filtering crucible with hot water. Remove any precipitate from the walls of the beaker with a rubber policeman and rinse such particles into the filter with hot water. If the filtrate is cloudy it must be refiltered, in which case the second passage generally clears it up. Continue to rinse the precipitate in the filter with hot water until a drop of silver nitrate solution added to a test portion of the washings collected in a test tube shows that chloride is absent.

After washing is complete, if filter paper was used, transfer the paper and precipitate carefully to the previously prepared crucible. Dry the precipitate slowly over a low flame (an oven may be used if desired). Then, increasing the heat, char the paper carefully and finally burn it off completely. Ignite the precipitate for about 20 min at the highest temperature of the Tirrill burner. The crucible should be uncovered and in a slanted position for free access of air to prevent reduction of barium sulfate to barium sulfide. (If such reduction is suspected, moisten the cooled precipitate with a little concentrated sulfuric acid and carefully raise the temperature, finishing the ignition once again at the highest temperature of the burner.) Cool and weigh the crucible and its contents, then reignite for a second 20-min period. This is repeated, of course, until constant weight is attained.

In the event that a filtering crucible was used instead of filter paper, ignition is the same except that there is no paper to be burned off; the crucible is heated inside an ordinary crucible, and a Meker burner is used. Be sure the precipitate is dried at a low temperature because steam formed upon strong ignition of wet material may carry precipitate out of the crucible.

Sulfur is usually reported as sulfur trioxide. The gravimetric factor in such a case is 0.3430. Calculate the percentage of sulfur trioxide in the unknown sample and report the results in the style required by the instructor.

Notes

1. Consult the instructor. Certain samples of hydrated salts should not be dried in the oven. The common samples purchased from commercial suppliers of student unknowns are usually to be dried.

2. Consult the instructor regarding the weight of sample.

IRON(III) HYDROXIDE

The gravimetric method for determining iron involves the precipitation of iron(III) hydroxide followed by ignition at high temperatures. The final weighing form is Fe_2O_3. This method is rarely used for iron in steel or alloys because of the convenient titrimetric methods available. In rock analyses, however, iron is separated from some other elements by precipitation of the hydroxide. Iron(III) hydroxide is a good example of a gelatinous precipitate, and hence the gravimetric determination of iron is often used as a student exercise.

Iron ores are commonly dissolved in hydrochloric acid, and then nitric acid or bromine is added to ensure oxidation of the iron to the $+3$ state. Ammonium hydroxide is added to bring about precipitation:

$$Fe^{3+} + 3NH_4OH \longrightarrow Fe(OH)_3(s) + 3NH_4^+$$

The precipitate does not have a composition corresponding to the formula $Fe(OH)_3$, but is more properly termed a *hydrous oxide*. The terms *hydroxide* and *hydrous oxide* are used interchangeably, however.

Properties

Iron(III) hydroxide is a very insoluble substance, the K_{sp} being about 10^{-36}. It is so insoluble that the $(Q - S)/S$ ratio is extremely large, and the initial particles are very small. These very small particles have a strong affinity for water, and when coagulation occurs, a large quantity of water is retained, resulting in a gelatinous precipitate. The precipitation occurs even in acid solution. The hydroxide is still gelatinous even when formed under conditions of increased solubility. Iron(II) hydroxide, $Fe(OH)_2$, is not quantitatively precipitated by ammonia in the presence of ammonium salts. For this reason, care must be taken to ensure that the iron is in the $+3$ state before precipitation is effected.

Coagulation of colloidal iron(III) hydroxide is aided by precipitation from a hot solution. The solution must not be boiled for a long time, however, since the gel will break up and the precipitate will become slimy. The precipitate is washed by decantation, using wash water that contains a small amount of ammonium nitrate. This salt prevents peptization of the precipitate and is preferred to ammonium chloride since the latter may form iron(III) chloride which is volatilized when the precipitate is ignited. Filtration is carried out using rapid filter paper. Suction is avoided because particles of the precipitate may clog the filter by being drawn into the pores. The filter paper is burned off and the precipitate ignited at fairly high temperatures since it tends to retain water rather strongly.

Errors

We have seen that gelatinous precipitates have a strong tendency to adsorb foreign ions from the solution during precipitation. By using some of the procedures described in Chapter 4 of the text, however, coprecipitation can be held to a minimum. If the precipitation is made from acid solution, the primary particles are positively charged since hydrogen ions are preferentially adsorbed. Hence cations are less strongly adsorbed in acid solution. Often higher *p*H values must be employed, however, and the particles are then negatively charged. Cations, particularly those that form insoluble hydroxides, are then strongly attracted by the precipitate. In such a case, it is recommended that a large amount of an ammonium salt be added to the solution so that the ammonium ion will be preferentially adsorbed. When the precipitate is ignited, the ammonium salt is volatilized and no error is made.

Since iron(III) hydroxide can be readily dissolved in acids, reprecipitation can be employed to advantage to rid the precipitate of adsorbed impurities. The precipitate is first thoroughly washed by decantation, leaving the bulk of the precipitate in the beaker. The precipitate is dissolved in hydrochloric acid and then re-formed by addition of ammonia.

Iron(III) phosphate and iron(III) arsenate are both precipitated in preference to iron(III) hydroxide. Hence the phosphate and arsenate ions cause errors if present in a sample. Silica is carried down by iron(III) hydroxide, causing high results. The ammonium hydroxide used to effect precipitation should be filtered to remove silica before use.

Iron(III) oxide is fairly easily reduced by carbon of the filter paper. Either the magnetic oxide, Fe_3O_4, or the metal may result. The filter paper should be burned off at a low temperature and a plentiful supply of air should be available during ignition in order to avoid this reduction. The iron(III) oxide can be formed again by treating the precipitate with concentrated nitric acid, drying it carefully, and reigniting.

Other Applications

Several other metals are precipitated as hydrous oxides by ammonia. Among these are aluminum, trivalent chromium, quadrivalent titanium, and quadrivalent manganese. The hydroxides of aluminum and chromium are amphoteric, and a large excess of ammonia must be avoided because these substances redissolve. Bivalent manganese is incompletely precipitated, but if an oxidizing agent such as bromine is added, hydrous manganese dioxide is precipitated. All of these hydrous oxides can be ignited to the oxides: Al_2O_3, Cr_2O_3, TiO_2, and Mn_3O_4. Thus the precipitation with ammonia can be used for determining each element provided the others are absent. However, the method is seldom used for manganese since the conversion of MnO_2 into Mn_3O_4 is not quantitative.

Precipitation with ammonia is used for the quantitative separation of iron and the foregoing elements from the alkali and alkaline earth cations.

EXPERIMENT 6.3. **Determination of Iron in an Oxide Ore**

Procedure

Accurately weigh out three samples of about 0.5 g each (Note 1) of the dried iron oxide unknown. Treat each sample in a clean 250-ml beaker with about 20 ml of 1 : 1 hydrochloric acid. Cover the beaker with a watch glass and heat the solution nearly to boiling. Continue this careful heating until solution of the sample is complete, or, if there is a persistent insoluble residue, until this residue (generally silica) shows no red-brown color.

Rinse down the watch glass and sides of the beaker with water from the wash bottle. Add 10 ml of saturated bromine water to the solution and boil gently until the bromine vapors are completely expelled. [If desired, the oxidation of iron(II) ion may be accomplished with nitric acid rather than bromine: add 1 to 2 ml of concentrated nitric acid dropwise to the solution and boil for a minute or two to expel the nitrogen oxides.]

If there is an undissolved siliceous residue in the sample, the solution should be filtered at this point. A "fast" paper is adequate. Collect the filtrate in a clean 400-ml beaker. Wash the filter paper with dilute hydrochloric acid until no yellow remains, collecting the washings in the same beaker.

Now dilute the solution to about 200 ml, bring it nearly to a boil, and slowly add 1 : 2 ammonium hydroxide solution until a definite odor of ammonia persists in the vapors (Note 2). Hold the solution at the near-boiling temperature for another minute or so and then allow the precipitate to settle. Test for completeness of precipitation with a drop or two of ammonium hydroxide.

The precipitate is now to be separated from the mother liquor, washed, redissolved, and reprecipitated. In the first filtration, decantation is used to the greatest extent possible, retaining the bulk of the precipitate in the original beaker for redissolving. Much time is saved by this technique. Carefully decant the supernatant solution through a "fast" filter paper (Whatman No. 41, for example), retaining the precipitate in the beaker as much as possible. Then wash this precipitate with about 25 ml of 0.1 M ammonium nitrate wash solution (8 or 10 g of NH_4NO_3/liter). Allow the precipitate to settle and decant the washings through the filter. Repeat this washing a second time. Discard the filtrate and washings, and place the beaker containing the precipitate under the funnel. Now pour through the funnel about 50 ml of 1 : 10 hydrochloric acid solution. This should completely dissolve any precipitate in the funnel and the bulk of the precipitate in the beaker. Make certain that the acid washes the filter paper thoroughly. This can be accomplished by adding the acid from a pipet, directing a slow stream near the upper edge of the paper all the way around its circumference. No trace of yellow should remain on the paper.

The iron is now reprecipitated with ammonium hydroxide exactly as before.

Allow the precipitate to settle, test for completeness of precipitation, and then decant the supernatant liquid through the same filter paper as used above after the first precipitation. Wash the precipitate by decantation three times with hot 0.1 *M* ammonium nitrate wash solution, finally bringing the precipitate onto the filter paper. Use a rubber policeman and rinsing to effect a quantitative transfer. (Alternatively, the inside of the beaker may be wiped with a small piece of ashless filter paper which is then added to the precipitate in the funnel.) The precipitate is now washed in the funnel with the hot ammonium nitrate solution until the filtrate shows only a faint test for chloride with silver nitrate.

After the precipitate has drained as much as possible, fold down the edges of the filter paper, and transfer paper with contents to a previously prepared porcelain crucible. Dry and char the paper in the usual manner and finally ignite the precipitate at bright red heat (a Meker burner is recommended, although a Tirrill burner may be adequate). There is some tendency for reduction of iron(III) oxide to the magnetic oxide, Fe_3O_4. Hence the ignition should be performed with plentiful access of air into the crucible. After ignition for about 45 min, cool the crucible in the desiccator and weigh. Reignite for 20-min periods until constant weight is attained.

Calculate the percentage of iron in your unknown sample. The gravimetric factor is 0.6994. Report the results as requested by the instructor.

Notes

1. Consult the instructor regarding recommended weight of sample.
2. Unless it was perfectly clear, the ammonium hydroxide should have been previously filtered to remove silica, which is often suspended in the solution as a result of attack on the glass container.

EXPERIMENT 6.4. Determination of Nickel in Steel

One of the most common applications of organic precipitants is the precipitation of nickel by dimethylglyoxime (Chapter 4 of the text). In the determination of nickel in steel, an acid solution containing iron in the +3 oxidation state is treated with tartaric acid and an alcoholic solution of dimethylglyoxime. The solution is made slightly basic with ammonia, precipitating the nickel quantitatively. Iron is not precipitated because it forms a soluble complex with tartrate ion.

Procedure

Weigh accurately three samples of about 1 g each (Note 1) of the steel into 400-ml beakers. Dissolve each sample by warming it with about 60 ml of 1:1 hydrochloric acid. Add cautiously 10 ml of 1:1 nitric acid and boil the solution gently to expel oxides of nitrogen. Dilute the solution to 200 ml, heat it nearly to boiling, and add 25 ml of a 20% solution of tartaric acid (Note 2).

Now neutralize the solution with concentrated ammonia until a definite odor of ammonia persists in the vapors. Then add 1 ml of excess ammonia. If any insoluble material is evident, remove it by filtration (filter paper), and wash it with a hot solution containing a little ammonia and ammonium chloride. Combine the washings with the remainder of the solution.

Make the solution slightly acidic with hydrochloric acid and heat it to about 70°C. Add 20 ml of a 1% solution of dimethylglyoxime in alcohol (Note 3). Make the solution slightly alkaline with ammonia (note odor) and add 1 ml of excess. Allow the precipitate to digest for 30 min at 60°C and then cool the solution to room temperature.

Filter the solution through a weighed fritted-glass or Gooch crucible, and wash the precipitate with cold water until the washings are free of chloride ion. Dry the precipitate by heating the crucible in an oven at 110 to 120°C until a constant weight is obtained. Calculate the percentage of nickel in the steel.

Notes

1. Consult the instructor. The sample should contain about 25 to 35 mg of nickel.

2. The solution is prepared by dissolving 25 g of tartaric acid in 100 ml of water. It should be filtered before use if it is not clear.

3. Dissolve 10 g of dimethylglyoxime in 1 liter of 95% ethanol. For best results, about 10 ml of this solution should be used for each 1% of nickel present. If too much solution is added, dimethylglyoxime itself may precipitate.

7

Optical Methods
of Analysis

The principles of spectrophotometry are discussed in Chapter 14 of the text and flame photometry in Chapter 15. The experiments given in this chapter are chosen to illustrate some of these principles as they are applied to chemical analysis and to the determination of physical constants.

EXPERIMENT 7.1. Determination of Manganese in Steel

Manganese in steel can be determined spectrophotometrically after oxidation to the purple permanganate ion.[1] The steel is dissolved in nitric acid and the oxidation is effected with potassium periodate:

$$2Mn^{2+} + 5IO_4^- + 3H_2O \longrightarrow 2MnO_4^- + 5IO_3^- + 6H^+$$

Since the yellow iron(III) ion is present, phosphoric acid is added to form the colorless iron-phosphate complex.

If the steel contains chromium or nickel, the color of these ions interferes in the manganese determination. This interference can be canceled by the addition of about the same amounts of these elements to the standards as are in the unknown. Alternatively, a sample can be carried through the entire procedure except for the periodate oxidation and then used as a blank in setting the spectrophotometer to read zero absorbance. In the directions below, a steel of known manganese content which contains about the same quantities of nickel and cobalt as the unknown is used as the standard.

The experiment first calls for determining a spectral-transmittance curve to

[1] H. H. Willard and L. H. Greathouse, *J. Am. Chem. Soc.*, **39**, 2366 (1917).

find the wavelength at which to perform the analysis. A check of Beer's law is then made by measuring the absorbances of permanganate solutions at several concentrations. The unknown is determined by comparison with the Beer's law plot.

Procedure

(a) *Spectral-Transmittance Curve.* Weigh accurately a sample of steel of known manganese content (Note 1). Dissolve the sample in 50 ml of 1:3 nitric acid, using heat and finally boiling gently for 1 to 2 min to remove oxides of nitrogen. Then remove the flask from the burner, add about 1 g of ammonium persulfate, and boil the solution gently for 10 to 15 min (Note 2). If a precipitate of an oxide of manganese forms or if the permanganate color develops, add a few drops of sodium sulfite or sulfurous acid (Note 3) and boil the solution a few minutes more to expel sulfur dioxide.

Dilute the solution with water to about 100 ml and add about 10 ml of 85% phosphoric acid and 0.5 g of potassium periodate. Boil the solution gently for about 3 min to effect oxidation to permanganate. Then cool the solution and dilute to 250 ml in a volumetric flask.

Measure the absorbance of the prepared solution or a suitable dilution thereof (Note 4) using the directions given for the spectrophotometer employed. Cover the range from about 440 to 700 nm,[2] taking readings at 20-nm intervals. In the region of maximum absorbance, take readings at intervals of 5 nm. Plot the absorbance vs. wavelength and connect the points to form a smooth curve. Select the proper wavelength to use for the determination of manganese (Note 5).

(b) *Beer's Law Check.* Secure four clean, dry 100-ml beakers or Erlenmeyer flasks and number them 1 to 4. In beaker 1 place some of the solution prepared in part (a) and do not dilute it. In beaker 2 place 30 ml of the standard solution (use a 10-ml pipet) and then add 10 ml of water. In beaker 3 place 10 ml of the standard solution and 10 ml of water, and in beaker 4 place 10 ml of the standard and 30 ml of water. Mix the solutions which were diluted and then measure the absorbance of each against distilled water at the wavelength of maximum absorbance. Plot the absorbance vs. concentration for the four solutions. Draw the best straight line between the points. Is Beer's law obeyed?

(c) *Analysis of Sample.* Weigh accurately a sample of steel whose manganese content is to be determined, dissolve it, and oxidize the manganese to permanganate as directed in part (a). The solution is finally diluted to 250 ml in a volumetric flask. Measure the absorbance of some of the solution against distilled water at the same wavelength used in part (b) to check Beer's law. Using the Beer's law plot, read off the concentration of permanganate and calculate the percentage of manganese in the sample.

---

[2] The abbreviation nm is for a *nanometer*, 10^{-9} m or 10^{-7} cm. It is synonymous with millimicron, mμ, a term no longer recommended.

Notes

1. Consult the instructor for the rough manganese content of the sample. Calculate the size sample needed to give a final solution whose absorbance falls in the range 0.7 to 0.8 if the molar absorptivity of permanganate is 2360 liters/mol-cm at 525 nm.

2. The ammonium persulfate is added to oxidize carbon or carbon compounds. The excess persulfate is destroyed by boiling the solution.

3. This reduces manganese to the bivalent state. The solution should be clear.

4. Any spectrophotometer or filter photometer can be used. Follow precisely the operating directions given by the instructor or in the operation manual.

5. This wavelength is 525 nm, but the wavelength calibration of the spectrophotometer is often not reliable and the value found with the instrument should be used.

EXPERIMENT 7.2. Determination of Iron with 1,10-Phenanthroline

The reaction between Fe^{2+} and 1,10-phenanthroline to form a red complex serves as a good sensitive method for determining iron. The molar absorptivity of the complex, $[(C_{12}H_8N_2)_3Fe]^{2+}$, is 11,100 at 508 nm. The intensity of the color is independent of pH in the range 2 to 9. The complex is very stable and the color intensity does not change appreciably over very long periods of time. Beer's law is obeyed.

The iron must be in the +2 oxidation state, and hence a reducing agent is added before the color is developed. Hydroxylamine, as its hydrochloride, can be used, the reaction being

$$2Fe^{3+} + 2NH_2OH + 2OH^- \longrightarrow 2Fe^{2+} + N_2 + 4H_2O$$

The pH is adjusted to a value between 6 and 9 by addition of ammonia or sodium acetate. An excellent discussion of interferences and of applications of this method is given by Sandell.[3]

Procedure[4]

Preparation of Solutions

(a) Dissolve 0.1 g of 1,10-phenanthroline monohydrate in 100 ml of distilled water, warming to effect solution if necessary.

(b) Dissolve 10 g of hydroxylamine hydrochloride in 100 ml of distilled water.

(c) Dissolve 10 g of sodium acetate in 100 ml of distilled water.

(d) Weigh accurately about 0.07 g of pure iron(II) ammonium sulfate, dissolve in water, and transfer the solution to a 1-liter volumetric flask. Add

[3] E. B. Sandell, *Colorimetric Determination of Traces of Metals*, 3rd ed., Interscience Publishers, Inc., New York, 1959.

[4] H. Diehl and G. F. Smith, *Quantitative Analysis*, John Wiley & Sons, Inc., New York, 1952

2.5 ml of concentrated sulfuric acid and dilute the solution to the mark. Calculate the concentration of the solution in mg of iron per ml.

Into five 100-ml volumetric flasks, pipet 1-, 5-, 10-, 25-, and 50-ml portions of the standard iron solution. Put 50 ml of distilled water in another flask to serve as the blank and a measured volume of unknown in another (Note). To each flask add 1 ml of the hydroxylamine solution, 10 ml of the 1,10-phenanthroline solution, and 8 ml of the sodium acetate solution. Then dilute all the solutions to the 100-ml marks and allow them to stand for 10 min.

Using the blank as the reference and any one of the iron solutions prepared above, measure the absorbance at different wavelengths in the interval 400 to 600 nm. Take readings about 20 nm apart except in the region of maximum absorbance where intervals of 5 nm are used. Plot the absorbance vs. wavelength and connect the points to form a smooth curve. Select the proper wavelength to use for the determination of iron with 1,10-phenanthroline.

Using the selected wavelength, measure the absorbance of each of the standard solutions and the unknown. Plot the absorbance vs. the concentration of the standards. Note whether Beer's law is obeyed. From the absorbance of the unknown solution, calculate the concentration of iron (mg/liter) in the original solution.

Note

Prepared solutions may be used as unknowns. Consult the instructor concerning size of sample to be used. If a natural water is used be sure that it is colorless and free of turbidity.

EXPERIMENT 7.3. Determination of Nitrite in Water[5]

The determination of nitrite ion in water is important in assessing the degree of pollution. The efficiency of a water purification process can be judged by the amount of nitrite ion in the water.

Nitrite ion can be determined in water by utilizing the reaction of this ion with amines (diazotization). The compound 4-aminobenzenesulfonic acid is diazotized according to the reaction.

$$HSO_3 - \langle \bigcirc \rangle - NH_2 + NO_2^- + 2H^+ \rightarrow HSO_3 - \langle \bigcirc \rangle - \overset{+}{N} \equiv N + 2H_2O$$

The diazonium salt is then coupled with 1-naphthylamine to form the colored product:

[5] M. G. Mellon, *Quantitative Analysis*, Thomas Y. Crowell Co., New York, 1955, p. 512. See also *Standard Methods for the Examination of Water and Sewage*, 10th ed., American Public Health Association, New York, 1955, p.153.

The solution is made slightly basic with sodium acetate in order to make this reaction complete.

Procedure

Preparation of Solutions

(a) Dissolve about 0.8 g of sulfanilic acid (4-aminobenzenesulfonic acid) in 28 ml of glacial acid and dilute the solution to about 100 ml with water.

(b) Dissolve about 0.5 g of 1-naphthylamine in 28 ml of glacial acetic acid and dilute the solution to about 100 ml with water.

(c) Dissolve about 14 g of sodium acetate trihydrate in water and dilute to about 50 ml.

(d) Weigh accurately 0.494 g of reagent-grade sodium nitrite, dissolve the salt in water, and dilute the solution to 1 liter in a volumetric flask. Pipet 10 ml of this solution into another 1-liter volumetric flask and dilute the solution to the mark. This solution now contains 0.0010 mg of nitrogen/ml.

Into seven 100-ml volumetric flasks, pipet 1-, 2-, 3-, 4-, 5-, 7-, and 10-ml portions of the standard nitrite solution. Secure two additional flasks, one for the blank and one for the unknown (Note 1). Pipet an aliquot (Note 2) of the unknown into one of these flasks. Then adjust the volume in each flask to about 50 ml with distilled water. Add to each flask 1 ml of the sulfanilic acid solution and allow the solutions to stand for 5 min. Then add to each flask 1 ml of the 1-naphthylamine solution and 1 ml of the sodium acetate. Finally, dilute each solution to the mark.

Using the blank as a reference and any of the nitrite solutions prepared above, measure the absorbance at different wavelengths in the interval 400 to 600 nm. Take readings about 20 nm apart except in the region of maximum absorbance, where intervals of 5 nm are used. Plot the absorbance vs. wavelength and connect the points to form a smooth curve. Select the proper wavelength to use for the determination.

Using the selected wavelength, measure the absorbance of each of the standard solutions and the unknown. Plot the absorbance vs. the concentration of the standards and note whether Beer's law is obeyed. From the absorbance of the unknown solution, calculate the number of milligrams of nitrogen per liter (ppm) of the original unknown solution.

Notes

1. Prepared solutions may be used as unknowns. If a natural water is used, be sure that it is colorless and free of turbidity.
2. Consult the instructor concerning the volume of sample to be used.

EXPERIMENT 7.4.   Determination of Phosphate in Water[6]

In acid solutions phosphoric acid reacts with molybdic acid to form a complex heteropoly acid, whose formula is sometimes represented as $H_3[P(Mo_3O_{10})_4]$. In aqueous solution this heteropoly acid has a yellow color which can be used as a basis for the colorimetric determination of phosphorus. Alternatively, the heteropoly acid can be reduced by a variety of reducing agents to give a blue solution, known as a heteropoly blue, or molybdenum blue, a compound of unknown composition. Solutions of molybdenum blue are more intensely colored than those of the yellow complex. Hence the colorimetric method based on molybdenum blue is more sensitive than the one based on the yellow complex.

The reaction which forms the yellow heteropoly acid is rapid in strongly acid solution. Molybdenum blue is formed at a slower rate, the reaction usually being complete within 10 to 15 min. The blue color of the reduced substance tends to fade because of side reactions.

The molybdenum blue method is sensitive to orthophosphate ions (PO_4^{3-}) and not to "condensed" phosphates such as $P_2O_7^{2-}$ and $P_2O_9^{3-}$. If an analysis for total phosphate is desired, the sample should be acidified and boiled for several minutes to convert polymeric phosphates to orthophosphates. The method is used to determine phosphate in the range 0.1 to 5 ppm. The major interference is the silicate ion, SiO_4^{4-}, which also forms a blue heteropoly product.

Procedure

Prepare a standard phosphate solution as follows. Weigh accurately 0.1834 g of dry K_2HPO_4, dissolve the sample in water, transfer the solution to a 1-liter volumetric flask, and dilute to the mark. Each milliliter contains 0.100 mg of PO_4, or 100 ppm. With a pipet transfer portions of 0.50, 1.0, 1.5, 2.0, and 2.5 ml of the standard phosphate solution to 100-ml volumetric flasks. Dilute each solution to about 50 ml with distilled water. Add about 50-ml of distilled water to another flask to serve as a blank. Secure a sample of the proper volume of the unknown

---

[6]See M. G. Mellon, *Quantitative Analysis*, Thomas Y. Crowell Co., New York, 1955, p. 515. The molybdenum blue method has been widely studied. The directions given here are based on studies of Stanley N. Deming of the University of Houston and are used with his permission.

from the instructor, place it in a 100-ml volumetric flask, and dilute to about 50 ml.

Treat each solution in the volumetric flasks as follows: Add about 2.0 ml of ammonium molybdate solution (Note 1) and swirl the flask several times to ensure thorough mixing. Then add 0.5 ml of $SnCl_2$ solution (Note 2) and swirl the flask again. Record the time at which the $SnCl_2$ is added, and then dilute the solution to the mark. Mix the solution thoroughly and allow the flask to stand for 10 min.

Ten minutes after the $SnCl_2$ was added (time zero), transfer some of the solution to a cuvette for measurement in the spectrophotometer (Note 3). About 11 min after time zero, place the cuvette in the spectrophotometer. At exactly 12 min after time zero, read and record the absorbance of the solution at 650 nm.

Repeat the foregoing procedure on the other standard solutions, the blank, and the unknown. Prepare a graph of absorbance vs. ppm of phosphate. Draw the best possible straight line through the points (Note 4), and determine the concentration of phosphate in your unknown solution. Report the ppm of phosphate in your unknown solution and turn in your graph to the instructor.

Notes

1. Dissolve 25 g of ammonium molybdate, $(NH_4)_6Mo_7O_{24} \cdot 4H_2O$, in about 175 ml of distilled water. Add to this a sulfuric acid solution prepared by mixing 310 ml of concentrated acid with 400 ml of water. Cool the solution to room temperature and dilute to 1 liter with water.

2. This is a 0.5 M solution prepared by dissolving 113 g of $SnCl_2 \cdot 2H_2O$ in 250 ml of concentrated hydrochloric acid, adding a few pieces of mossy tin, and diluting to 1 liter with water.

3. The color of molybdenum blue gradually fades. Reproducible results can be obtained, however, if the absorbance readings are taken at the same time interval after mixing the solutions.

4. There may be some deviation from Beer's law above about 1.5 ppm. If some curvature is apparent, draw a smooth curve through the points.

EXPERIMENT 7.5.   Determination of Glucose[7]

Glucose, also known as dextrose, is a simple sugar with the empirical formula $C_6H_{12}O_6$. It forms important polymers such as starch, cellulose, and glycogen. Carbohydrates are absorbed in the bloodstream as glucose and the sugar is oxidized by the body to produce energy.

The concentration of glucose in the blood is of interest medically. A deficiency of the hormone *insulin*, which promotes the metabolism of glucose, leads to the disease called *diabetes*. Insufficient insulin leads to an increase in blood glucose concentration and changes in fat metabolism, which can cause ketosis and

---

[7] This experiment was designed by Ronald C. Johnson of Emory University, who has kindly given us permission to use it here.

possible diabetic coma. An excess of insulin causes blood glucose levels to become abnormally low and can lead to "insulin shock."

A rapid and reliable measurement of the concentration of glucose is very important in clinical chemistry. A number of methods are available. In this experiment we shall use the method based on the oxidation of glucose by ferricyanide ion, $Fe(CN)_6^{3-}$. The method is well suited for the automated analyzers used in most clinical laboratories today.

In hot alkaline solution yellow ferricyanide ion is reduced by glucose to colorless ferrocyanide ion:

$$Fe(CN)_6^{3-} + e \longrightarrow Fe(CN)_6^{4-}$$

$$\text{yellow} \qquad\qquad\qquad \text{colorless}$$

The absorbance of the ferricyanide ion is measured at 420 nm. In solutions containing glucose, ferricyanide is reduced and the absorbance decreases. Standard solutions are prepared which initially contain equal concentrations of ferricyanide, but varying amounts of glucose. After reaction has occurred the absorbance of each solution is measured and is plotted against the initial concentration of glucose. The decrease in absorbance is proportional to the concentration of glucose.

The experiment uses aqueous glucose solutions rather than actual blood samples. The concentrations of glucose are about one-tenth that found in the blood. The 5-ml sample size is much larger than that used in clinical laboratories, where autoanalyzers make it convenient to use much smaller blood samples.

Procedure

Your instructor will furnish two stock solutions:
  (a) 0.015 M in $Fe(CN)_6^{3-}$; 0.50 M in Na_2CO_3.
  (b) Glucose: 5.000 mg/100 ml.

Introduce 2.00 ml of solution A into each of five clean 50-ml volumetric flasks using a pipet. Mark the flasks 1, 2, 3, 4, and U (for unknown). Now pipet 5.00 ml of solution B into volumetric flask 2; 10.00 ml of solution B into flask 3; and 15.00 ml of B into flask 4. Obtain an unknown solution from your instructor and pipet 5.00 ml of it into the flask marked U. Record the number of the unknown.

To each of the five volumetric flasks add distilled water, rinsing down the walls, until the flasks are about half-filled. Place water in a small metal pan and heat it to boiling. Place the *uncapped* volumetric flasks in the boiling water for 15 min, being careful that the flasks do not capsize. Cool the flasks to room temperature using tap water, and then dilute each solution to the mark using distilled water. Cap the flasks and mix the contents thoroughly by inverting the flasks several times.

Using distilled water as the reference, measure the absorbance of each of the standard solutions and the unknown at 420 nm. Plot the absorbance vs. the concentrations of the original glucose solutions (i.e., 0.00, 5.00, 10.0, and 15.0 mg/ml, respectively). Draw the best straight line through the points. Determine the concentration of glucose in your unknown solution using this graph, and report your results as directed to the instructor.

EXPERIMENT 7.6. Analysis of a Permanganate–Dichromate Mixture

As described in Chapter 14 of the text, it is sometimes possible to determine more than one constituent in a solution using spectrophotometric measurements. This experiment illustrates the procedure. Permanganate and dichromate ions both absorb visible light, but the absorbance peaks are fairly well separated. By measuring the absorbance at two different wavelengths, it is possible to determine the concentrations of the two ions in the same solution. Lingane and Collat[8] have studied these two systems thoroughly in applying the method to the determination of manganese and chromium in steel.

Procedure

(a) Permanganate Standards. Prepare and standardize a 0.01 M solution of potassium permanganate as directed in Experiments 4.1 and 4.2. Then, using a measuring pipet, place 0.50-, 1.00-, 2.00-, 3.00-, and 4.00-ml portions into five 100-ml volumetric flasks numbered 1 to 5. Dilute the solution in each flask to the mark with 0.25 M H_2SO_4.

(b) Dichromate Standards. Prepare 100 ml of about 0.02 M $K_2Cr_2O_7$ by weighing 0.55 to 0.60 g of the reagent-grade substance on the analytical balance, dissolving in water, and diluting to 100 ml in a volumetric flask. Then, using a measuring pipet, place 2.00-, 4.00-, 6.00-, 8.00- and 10.00-ml portions into five 100-ml volumetric flasks numbered 1 to 5. Dilute the solution in each flask to the mark with 0.25 M H_2SO_4.

(c) Absorption Spectra and Beer's Law Check. Using one of the permanganate solutions of intermediate concentration, measure the absorbance over the wavelength interval 350 to 650 nm using 0.25 M H_2SO_4 solution as the reference. Read the absorbance at 20-nm intervals except in the vicinity of the maxima and minima in the spectrum, where readings should be taken every 5 nm. Plot absorbance vs. wavelength. Repeat the procedure for one of the dichromate solutions.

Select two wavelengths for the determination (Note 1). Then measure the absorbances of the standard solutions at each wavelength. Plot absorbance vs. concentration for each ion at the two wavelengths and calculate the molar absorptivity (Note 2) for each substance at each wavelength.

(d) Analysis of Unknown. Secure an unknown solution from the instructor and dilute it according to his instructions. Then measure the absorbance at the two

[8]J. J. Lingane and J. W. Collat, *Anal. Chem.*, **22**, 166 (1950).

chosen wavelengths. Calculate the concentration of MnO_4^- and of $Cr_2O_7^{2-}$ in the unknown and report this to the instructor.

Notes

1. Lingane and Collat[9] suggest 440 and 545 nm for iron samples. Students should determine the best values for their use with the simple mixture and their own spectrophotometer.

2. If the path length is not known, simply calculate ϵb for each wavelength.

EXPERIMENT 7.7.   **Spectrophotometric Determination of the pK_a of an Acid-Base Indicator[10]**

In this experiment, spectrophotometry is employed to measure the pK_a of bromthymol blue, an acid-base indicator. The indicator (HIn) is a monobasic acid and we can represent its dissociation as follows:

$$HIn \rightleftharpoons \; + In^-$$

As is shown in Chapter 5 of the text, the equilibrium expression for such a dissociation can be written

$$pH = pK_a - \log \frac{[HIn]}{[In^-]}$$

This can be rearranged to give

$$\log \frac{[In^-]}{[HIn]} = pH - pK_a$$

This is in the slope-intercept form of an equation for a straight line,

$$y = mx + b$$

where y is $\log [In^-]/[HIn]$, $m = 1$ and $b = -pK_a$. Hence, if the log term is plotted vs. pH, the slope is 1, the intercept is $-pK_a$, and the line should cross the pH axis at $pH = pK_a$ (Fig. 7.1). At the latter point $[In^-] = [HIn]$, and hence the log of the ratio of these terms is zero, making $pH = pK_a$.

The ratio $[In^-]/[HIn]$ can be determined spectrophotometrically. First, a solution of bromthymol blue is prepared in acidic solution (low pH) where essentially all of the indicator is in the HIn form. The absorption spectrum is then determined. Second, a solution is prepared in basic solution (high pH) where essentially all of the indicator is in the In^- form. The absorption spectrum of In^- is then determined. From the two absorption spectra the wavelengths of maximum absorbance of HIn and In^- are selected for further measurements.

[9] *Ibid.*, p. 166.

[10] C. N. Reilley and D. T. Sawyer, *Experiments for Instrumental Methods*, McGraw-Hill Book Company, New York, 1961.

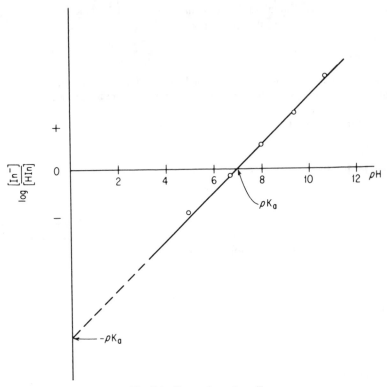

Fig. 7.1 Plot to determine pK_a.

Buffered solutions with pH values on either side of the pK_a of bromthymol blue are then prepared and the absorbances measured at the selected wavelengths. The solutions contain the same total concentration of indicator, $\{[HIn] + [In^-]\}$, but the ratios vary with pH. Figure 7.2 shows a typical plot of absorbance vs. pH at the wavelength of maximum absorbance for the In^- species. The terms used in this figure are as follows:

$$A_a = \text{absorbance of HIn}$$

$$A_b = \text{absorbance of In}^-$$

$$A = \text{absorbance of mixture}$$

From the graph it is evident that

$$\frac{[In^-]}{[HIn]} = \frac{A - A_a}{A_b - A}$$

If the wavelength used is the one at which HIn shows maximum absorbance, the curve will be similar to that shown in Fig. 7.2 except that it will start at a high absorbance and curve down to a low absorbance value at high pH.

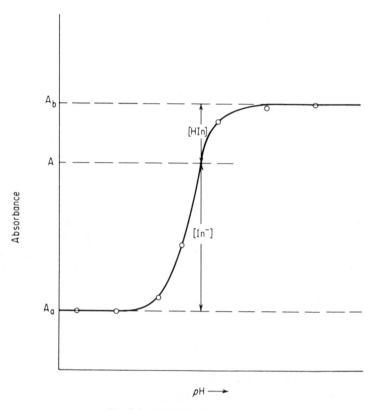

Fig. 7.2 Plot of absorbances vs. *p*H.

Procedure

Preparation of Solutions

(a) Dissolve 0.1 g of bromthymol blue in 100 ml of 20% ethanol.

(b) Dissolve 2.4 g of NaH_2PO_4 in water and dilute to 100 ml. The solution is 0.2 *M*.

(c) Dissolve 3.4 g of K_2HPO_4 in water and dilute to 100 ml. The solution is 0.2 *M*.

(d) Prepare a small amount of 3 *M* NaOH. Three grams of NaOH dissolved in 25 ml of solution will provide more than enough of this reagent.

Prepare a series of buffered solutions as follows: Secure seven clean 50-ml volumetric flasks (Note 1). Place 2.0 ml of the bromthymol blue solution in each flask, using a pipet. Then add the volume of phosphate solutions shown in the chart to these flasks. (These volumes can be measured with a graduated cylinder.) To flask 7 add 2 drops of the 3 *M* NaOH solution. Now dilute each solution to the mark and mix thoroughly. Measure and record the pH value of each solution using a pH meter.

Flask	NaH$_2$PO$_4$, ml	K$_2$HPO$_4$, ml
1	5	0
2	5	1
3	10	5
4	5	10
5	1	5
6	1	10
7	0	5

To determine the absorption spectrum of bromthymol blue at low pH (the spectrum of HIn), measure the absorbance of solution 1 from 400 to 640 nm, using water as a reference (Note 2). Read the absorbance at 20-nm intervals except in the vicinity of the maximum in the spectrum, where readings should be taken every 5 nm. Plot the absorbance vs. wavelength.

To determine the absorption spectrum at high pH (the spectrum of In$^-$), measure the absorbance solution 7 in the same manner as directed above. Plot the results on the same piece of graph paper.

Using the spectra obtained above, select two wavelengths at which further absorbance measurements will be made. Choose wavelengths at which HIn and In$^-$ exhibit maximal differences in absorbance.

Now measure the absorbances of each of the seven solutions at the two wavelengths selected (Note 3). Prepare a graph of absorbance vs. pH for each of the two wavelengths. (See Fig. 7.2) Determine the [In$^-$]/[HIn] ratios in solutions 2 to 6 as explained above and shown in Fig. 7.2. Plot log [In$^-$]/[HIn] vs. pH (Fig. 7.1) and obtain a value of pK_a from the graph (Note 4).

Report the value of the pK_a of bromthymol blue in the manner desired by the instructor. Turn in all your graphs.

Notes

1. If it is more convenient to use 100-ml or 25-ml volumetric flasks, the amounts of reagents should be scaled up or down proportionately.

2. The pH of this solution is sufficiently low that most of the indicator is in the HIn form. If the absorbance reading at any wavelength is very high (above 0.8 and 1.0) the solutions should be diluted or less bromthymol blue used in each flask.

3. Before making the measurements, consult the instructor, who may wish measurements made at one wavelength only.

4. Note that the value of pK_a can also be obtained from the graph in Fig. 7.2 by reading the pH at which $A_b - A = A_a - A$ (i.e., the absorbance halfway between A_a and A_b). Consult the instructor, who may wish you to determine the pK_a by several methods and report the average.

EXPERIMENT 7.8.   **Determination of the Formula of a Complex Ion** [11]

In this experiment the formula of the complex ion formed in solution from copper(II) and EDTA is determined using spectrophotometric measurements. The methods employed are widely used in inorganic and analytical chemistry.

Consider the case where a cation M^{2+} reacts with a ligand X to form a colored complex MX_n^{2+}:

$$M^{2+} + nX \rightleftharpoons MX_n^{2+}$$

An example is the reaction of copper(II) ions with molecules of ammonia to form species such as $Cu(NH_3)_4^{2+}$. Let us assume that M^{2+} and X are colorless and that the complex MX_n^{2+} is colored. The problem is to evaluate n to determine the formula of the complex.

The method of continuous variations involves measuring the absorbances of a series of solutions of varying composition at a wavelength at which the complex shows its maximum absorptivity. The total number of moles of M^{2+} and X is kept constant while the mole fractions of the reactants are varied. The absorbances are plotted against mole fraction and the value of n can be determined from the mole fraction at which maximum absorbance occurs. In Fig. 7.3 an illustration is given in which the peak occurs at a mole fraction of 0.75. This indicates that the formula of the complex is MX_3^{2+}. The graph is curved near the point of maximum absorbance as shown in Figure 7.3. The departure from linearity is greater the less stable the complex. The straight-line portions of the graph are extrapolated until they cross to determine the mole fraction at maximal absorbance.

The mole-ratio method employs a series of solutions in which the concentration of one reactant, say M^{2+}, is constant and that of the other, say X, is varied. The absorbances are plotted against the ratio of the moles of ligand to metal, as shown in Fig. 7.4. The absorbance increases to a maximal value at a mole ratio of 3:1 and then levels off. Again, the graph may be curved near the peak and the straight-line portions may be extrapolated until they intersect.

The mole-ratio method and the method of continuous variations are most readily applied when only a single complex is formed between the reacting species. Copper(II) and EDTA react to form a deep blue 1:1 complex which shows a flat maximum in absorbance in the range 730 to 755 nm. The reaction in a chloroacetate buffer (chloroacetic acid = HX) can be written

$$CuX_2 + H_2Y^{2-} \rightleftharpoons CuY^{2-} + 2HX$$

blue green   deep blue

The copper is present as the chloroacetate complex in the buffer and the chloroacetate ligand is displaced by EDTA, since the latter forms a much more stable complex with copper ion.

---

[11] This experiment was suggested by Hubert L. Youmans and has been checked by his students at Western Carolina University.

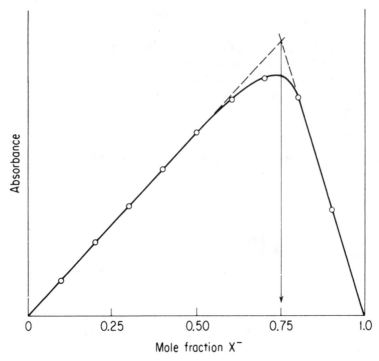

Fig. 7.3 Continuous-variations plot.

Procedure

Preparation of Solutions

(a) *Chloroacetate buffer.* Dissolve 18.2 g of chloroacetic acid in 100 ml of water and adjust the pH to about 2 using 5 *M* NaOH. Then dilute the solution to 250 ml.

(b) *Cu^{2+}, 0.0200 M.* Weigh accurately a sample of clean copper wire or foil of about 0.32 g (0.3175 g for exactly 0.0200 *M*), and place it in a 250-ml beaker. Add 5 ml of 1:1 nitric acid and dissolve the copper by warming the solution on a steam bath or over a low flame. Add 25 ml of water and boil the solution for about 1 min. Then add 5 ml of urea solution (1 g in 20 ml of water) and continue boiling for another minute to remove oxides of nitrogen. Cool the solution in tap water and neutralize the acid with 1:3 ammonia, adding the ammonia carefully until a pale blue precipitate of copper(II) hydroxide is obtained. Now add glacial acetic acid drop by drop until the precipitate redissolves. Transfer the solution to a 250-ml volumetric flask, dilute to the mark, and mix by inverting the flask.

(c) *EDTA, 0.0200 M.* Dissolve 1.861 g of disodium dihydrogen EDTA di-

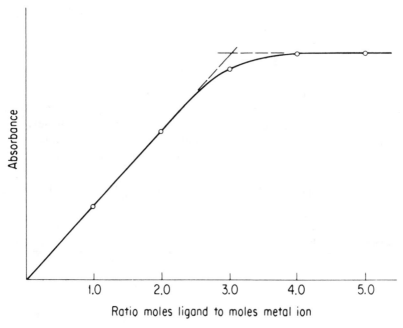

Fig. 7.4 Mole-ratio plot.

hydrate in about 100 ml of water in a 250-ml beaker. Transfer the solution to a 250-ml volumetric flask and dilute to the mark. Mix the solution thoroughly.

Method of Continuous Variations. Using a 10-ml measuring pipet, prepare the following mixtures from the preceding solutions.

Solution	Buffer, ml	Cu^{2+}, ml	EDTA, ml
1	10.0	10.0	0.0
2	10.0	9.0	1.0
3	10.0	7.0	3.0
4	10.0	5.0	5.0
5	10.0	3.0	7.0
6	10.0	1.0	9.0
7	10.0	0.0	10.0

Stir the solutions and measure the absorbance of each at 730 nm using distilled water as the reference. For the measurements at 730 nm the infrared-sensitive phototube and a red filter should be used.

Since the absorbance of the chloroacetate complex is appreciable (as shown by solution 1), the absorbances measured must be corrected to give the absorbance due to the copper-EDTA complex alone. This correction should be made for solutions 1, 2, and 3 as follows:

$$A_c = A_m - \frac{\text{ml Cu}^{2+} - \text{ml EDTA}}{10} \times A_1$$

where A_m is the absorbance measured, A_1 is the absorbance of solution 1, and A_c is the corrected absorbance. No correction need be made for solutions in which sufficient EDTA has been added to convert all the copper to the complex.

Plot the corrected absorbances vs. mole fraction EDTA. From the peak in the graph determine the formula of the complex and report this to the instructor.

Mole-Ratio Method. Prepare the following mixtures from the prepared solutions.

Solution	Buffer, ml	Cu^{2+}, ml	EDTA, ml	H_2O, ml
1	5.0	10.0	0.0	20.0
2	5.0	10.0	2.0	18.0
3	5.0	10.0	5.0	15.0
4	5.0	10.0	8.0	12.0
5	5.0	10.0	10.0	10.0
6	5.0	10.0	15.0	5.0
7	5.0	10.0	20.0	0.0

Stir the solutions and measure the absorbance of each at 730 nm using distilled water as the reference. The absorbances of solutions 1, 2, 3, and 4 should be corrected as follows:

$$A_c = A_m - \frac{10.0 - \text{ml EDTA}}{10.0} \times A_1$$

where A_m is the absorbance measured, A_1 is the absorbance of solution 1, and A_c is the corrected absorbance. No correction need be made for the other solutions.

Plot the corrected absorbances against the ratio of the volume of EDTA to the volume of Cu^{2+} solution. Determine the formula of the complex and report this to the instructor.

EXPERIMENT 7.9. **Photometric Titration of Bismuth and Copper with EDTA**[12]

The principles of photometric titrations are discussed in Chapter 14 of the text. In this experiment bismuth is titrated with the chelating agent ethylenediaminetetraacetic acid using copper(II) ion at a wavelength where the copper(II)EDTA chelate

[12] A. L. Underwood, *Anal. Chem.*, **26**, 1322 (1954).

absorbs strongly and the other species, Bi^{3+}, bismuth-EDTA chelate, and EDTA do not absorb. As the EDTA titrant is added to a Bi^{3+}-Cu^{2+} mixture, the Bi^{3+} reacts first since the bismuth chelate is much more stable than the copper(II) chelate. The absorbance remains constant (zero) until the bismuth ion concentration is reduced to a very low value. It then begins to rise as the copper(II)-EDTA chelate is formed. After the copper has reacted, the absorbance levels off again as excess nonabsorbing EDTA is added.

The titration is carried out at a *p*H of about 2.0. At lower *p*H values the end point is not as sharp, and above *p*H 2.5 there is danger of precipitating bismuth.

Procedure

Preparation of Solutions

(a) *EDTA, 0.1 M.* Dissolve about 3.8 g of disodium dihydrogen EDTA dihydrate in 100 ml of solution. The solution will be standardized by titration against pure bismuth.

(b) *Cu^{2+}, 0.2 M.* Dissolve about 2.4 g of $Cu(NO_3)_2 \cdot 3H_2O$ in 50 ml of solution.

(c) *NaOH, 5 M.* Dissolve about 10 g of NaOH in 50 ml of solution.

(d) *Bi^{3+}, 0.01 M.* Weigh accurately about 0.2 g of pure bismuth metal and dissolve it in the minimal quantity of 1 : 1 nitric acid using gentle heating. Transfer the solution to a 100-ml volumetric flask and dilute to the mark with distilled water which contains sufficient nitric acid to make the final solution about 0.5 *M* in acid.

Pipet a 20-ml aliquot of standard bismuth solution into a 250-ml beaker. Add 2 g of solid chloroacetic acid (Note 1) and 1 ml of the copper(II) nitrate solution. Adjust the volume to about 100 ml and adjust the *p*H to about 2, using 5 *M* NaOH. Read the *p*H with a *p*H meter. Pour carefully some of this solution into a cuvette; place the cuvette in the spectrophotometer and set the instrument to read zero absorbance at 745 nm (Note 2). Return the solution to the beaker and add from a 10-ml buret about 0.4 ml of the EDTA solution. Mix the solution thoroughly and rinse the cuvette with the solution, returning the rinsings to the beaker. Then fill the cuvette and measure the absorbance again. Repeat this procedure, adding the titrant in about 0.4-ml increments until the absorbance rises and finally reaches a constant value again. It is advisable to check the instrument occasionally in case of drift. This can be done using some of the untitrated bismuth solution in a separate cuvette.

Plot the absorbance readings (Note 3) against milliliters of titrant, draw straight lines through the points, and read the end-point volume for the first end point from the intersection of the two lines. The second end point will not be used in this case. Calculate the molarity of the EDTA solution.

Secure from the instructor a solution containing bismuth and copper of unknown concentrations. Treat the solution exactly as in the standardization procedure above except omit the addition of copper(II) nitrate solution. Titrate with EDTA as directed above and calculate the molarities of both bismuth and copper. Report these values to the instructor.

Notes

1. A chloroacetate buffer is appropriate because of the *pK* value of chloroacetic acid. If the acid is discolored it should be distilled before use.

2. For measurements at 745 nm the infrared sensitive phototube and a red filter should be used. Consult the instruction manual for directions.

3. The absorbance values should be corrected for dilution by the titrant. This will probably have little effect on the end-point volume, however. Consult the instructor.

EXPERIMENT 7.10. Ultraviolet Analysis of Benzene in Cyclohexane[13]

Spectrophotometry in the near-ultraviolet region (200 to 390 nm) can be applied to the qualitative and quantitative determination of many organic and some inorganic compounds (Chapter 14). Such compounds as aldehydes, ketones, aliphatic nitro compounds, and nitrate esters absorb in this region, although the intensities are so low that the spectra are useful only under special circumstances. Molecules containing conjugated double bonds, such as benzene and toluene, have rather high molar absorbances in the near ultraviolet and can be determined spectrophotometrically. Other molecules which show strong absorption in this region include azo, diazo, and nitroso compounds, quinones, and nitrile esters.

This experiment involves the spectrophotometric determination of benzene in cyclohexane. The differential method is employed. A recording spectrophotometer is desirable for making the measurements.

Procedure

Spectrum of Benzene Vapor. Record a spectrum of benzene vapor against air in the reference cell from 220 to 290 nm. To obtain the sample of benzene vapor, carefully put 1 drop of benzene in a cuvette. Do not get any liquid benzene on the side of the cuvette. Cap the cuvette, wait about 2 min for liquid-vapor equilibrium to be established, then record the spectrum.

Preparation of Standards. Secure a stock solution of 10.0 v/v% benzene in cyclohexane from the instructor. With this solution prepare six standards in the range 0.2 to 1.0%. Use a 10-ml buret with a Teflon stopcock to measure the stock solution into 25-ml volumetric flasks. Dilute to volume with spectrographic-grade cyclohexane. Use only solvents furnished by the instructor (Note 1).

---

[13] This experiment was designed by Hubert L. Youmans of Western Carolina University, who has kindly given us permission to use it here.

Standard Curve and Analysis. Record a spectrum of the 1% standard against the 0.2% standard as the reference. Choose a wavelength where the absorbance falls between 0.7 and 1.0. Then measure the absorbance of all the standard solutions at this wavelength. Take a 25-ml volumetric flask to your instructor for an unknown sample. Dilute it to volume with cyclohexane and measure its absorbance at the wavelength used for the standards (Note 2). Prepare a calibration curve and use it to determine the v/v% benzene in the unknown after it has been diluted to volume.

Explain the difference between the spectra of benzene vapor and benzene in cyclohexane.

Notes

1. These standards can also be used in Experiments 7.11 and 10.3.
2. The unknowns will contain 0.2 to 1.0% benzene when diluted to volume. The instructor may give more than one sample for analysis.

EXPERIMENT 7.11.   Infrared Analysis of Benzene in Cyclohexane[14]

The use of infrared spectrophotometry in analytical chemistry is discussed in Chapter 14. This experiment involves the determination of benzene in cyclohexane using an infrared spectrophotometer.

Direct application of Beer's law is very difficult in the infrared region because of instrumental limitations. The source and detector have rather limited output. The instrument thus uses a rather large slit width and a wide band of wavelengths, whereas most infrared spectral peaks are rather narrow. For this reason most infrared spectrophotometers cannot measure the actual height of the peak chosen for analysis but measure an average or integrated height across the peak in the range of wavelengths passed by the instrument. Another instrumental factor that produces deviation from Beer's law is the large amount of stray radiation.

A calibration curve must be prepared and checked frequently for direct quantitative work in the infrared. A baseline technique is generally used. An internal standard method is also frequently employed.

Procedure

Secure a stock solution of 10.0 v/v% benzene in cyclohexane from the instructor. With this solution prepare six standards in the range 0.2 to 1.0% (Note 1). Use a 10-ml buret with a Teflon stopcock to measure the stock solution into 25-ml volumetric flasks. Dilute to volume with spectrographic-grade cyclohexane.

Record spectra of 0.2% and 1.0% benzene in cyclohexane over the full range of the instrument. Use the attenuator in the reference beam. Start each spectrum

---

[14]This experiment was designed by Hubert L. Youmans of Western Carolina University, who has kindly given us permission to use it here.

with the pen set at 90% *T* and scan at fast speed. Compare the spectra and determine a band suitable for quantitative analysis. There is really only one.

Record spectra of the standards and unknown (Note 2). You need record only a small portion of the spectra near the transmittance minimum of the analytical band. Set 100% transmittance in a flat portion of the spectrum and record through the transmittance minimum. Start each time at the same wavelength. You should be able to record all spectra on one sheet of recorder paper by shifting the drum. Use a slow scan speed.

Plot %*T* vs. concentration on semilog paper or log %*T* vs. concentration on regular graph paper. Connect the points with an appropriate curve or straight line and determine the concentration of your unknown from this graph.

Notes

1. The standards prepared in Experiment 7.10 can be used here.
2. Take a 25-ml volumetric flask to your instructor for an unknown sample and dilute it to volume with cyclohexane.

EXPERIMENT 7.12.   Determination of Copper by Atomic Absorption Spectroscopy[15]

The principles of atomic absorption spectroscopy are discussed in Chapter 14 of the text. Your instructor will furnish directions on the operation of the atomic absorption spectrophotometer you will use. The experiment involves the determination of copper in Monel, an alloy from which nickel coins are made. It contains 20 to 30% copper.

Procedure

Copper shot is used as the primary standard. Weigh on the analytical balance about 0.3 g of copper shot and dissolve it in 20 ml of 1 : 1 water–nitric acid. After dissolving the alloy, transfer the solution to a 1-liter volumetric flask and dilute it to volume with distilled water.

Pipet 0, 2, 4, 5, 7, 9, and 10 ml of the standard solution into a series of 100-ml volumetric flasks. To each flask add 10 ml of 0.2% Sterox solution with a pipet. Sterox is a nonionic detergent. Dilute the solutions to volume with deionized water and mix. Express the concentrations as ppm Cu.

Weigh duplicate samples of the monel unknown on the analytical balance and dissolve each in 20 ml of 1 : 1 water–nitric acid. Dilute each sample to 1 liter in a volumetric flask. Pipet 10 ml of each solution and 10 ml of 0.2% Sterox into 100-ml volumetric flasks and dilute to volume with deionized water.

[15] This experiment was designed by Hubert L. Youmans of Western Carolina University, who has kindly given us permission to use it here.

The atomic absorption spectrophotometer is operated in the absorption mode. Set 100% T with the blank and measure %T of the standards and unknown. Plot %T vs. concentration on semilog paper or convert %T to absorbance and plot A vs. concentration on linear paper. Report the %Cu in the Monel alloy.

EXPERIMENT 7.13. Flame Photometric Determination of Calcium in Water[16]

The principles of flame photometry are discussed in Chapter 15 of the text. Your instructor will furnish directions on the operation of the flame photometer you will use. The experiment involves the determination of calcium in water. The method of standard addition is used. In this method known amounts of a standard solution of the analyte are added to aliquots of the unknown. Each aliquot of unknown is of the same size but the amount of standard is varied. The emission intensity is plotted against the concentration of standard (right side of ordinate). If the points fall on a straight line (of course there may be some "scatter"), the line may be extrapolated to the abscissa on the left side of the ordinate. The intercept gives the concentration of the unknown. If the right side of the line is curved, the standard addition method should not be used.

Procedure

All solutions should be made with deionized water. Distilled water contains a measurable amount of metal ions. Glassware should be rinsed with deionized water.

Prepare a standard solution containing 10.0 ppm of Ca^{2+} by dissolving 0.250 g of dried $CaCO_3$ in 5 ml of 6 M HNO_3. Dilute this solution to 1-liter in a volumetric flask; then pipet 10 ml of this into a 100-ml volumetric flask and dilute to the mark.

Number five 100-ml volumetric flasks B, 0, 5, 10, and 15. To the flasks pipet in order 0, 0, 5, 10, and 15 ml of the 10.0 ppm Ca standard. To each flask pipet 10 ml of 0.2% Sterox solution. To all flasks except B pipet 10 ml of the water sample. Dilute each flask to volume with deionized water and then mix.

Measure the emission intensities and make a standard addition method graph. Set 0 emission with solution B. Determine if the instrument operates linearly over the range of the analysis. Determine the ppm (w/v basis) of Ca in the unknown. Report the concentration to three significant figures.

---

[16]This experiment was designed by Hubert L. Youmans of Western Carolina University, who has kindly given us permission to use it here.

8

Potentiometric Titrations

It was pointed out in Chapter 12 of the text that a potentiometric titration involves measurement of the difference in potential between an indicator electrode and a reference electrode during a titration. The difference in potential can be measured with an ordinary potentiometer or with a pH meter. Generally, precise measurements of potential differences are made with a potentiometer. However, for the precision required in titrations a pH meter gives satisfactory results and is frequently more convenient to use. A brief discussion of the potentiometer and pH meter is given below, followed by directions for performing several experiments.

Potentiometer

A diagram of a potentiometer circuit is shown in Fig. 8.1. The potential of the unknown cell X is determined by measuring the potential which must be applied in opposition to X in order to prevent a flow of current through the galvanometer G. The working cell is a battery B, which applies a potential to the slide wire AD. This wire is a conductor of uniform resistance, and hence the fall in potential between any two points on the wire is uniform. The total potential drop along the wire is controlled by the rheostat R. The contact C is moved along the slide wire, and the key K is tapped until the galvanometer shows no deflection.

It is necessary to calibrate the slide wire, (e.g., to establish a known potential difference between A and D) in order to determine the potential of the unknown cell. This is done by replacing the cell X with a cell whose potential is accurately known—a *standard cell*. The most widely used standard cell is one of mercury and cadmium, called the *Weston cell*. Its potential at 25°C is 1.0186 V. The slide wire

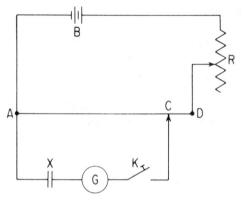

Fig. 8.1 Simple potentiometer circuit.

is stretched along a scale which is graduated uniformly. For example, if the scale has 1200 divisions from A to D, the sliding contact C is set at 1018.6 divisions from A, and the rheostat is adjusted until the galvanometer deflection is zero. Then the potential difference between A and C is 1.0186 V, and the fall in potential per scale division is 1 mV.

pH Meters

An ordinary potentiometer cannot be employed with a glass electrode because of the high resistance, 1 to 100 MΩ, of this electrode. The pH meter is a voltage measuring device designed to be used with cells of high resistance. There are two common types available commercially, the potentiometric and the direct-reading. The former is basically a potentiometer, but since the off-balance currents are so small because of the high resistance of the cell, the current is amplified electronically so that it will affect a galvanometer or microammeter. The direct-reading instruments are electronic voltmeters of very high input resistance; the circuit is so arranged as to give a meter reading proportional to pH. The voltage of the glass-reference electrode pair is impressed across a very high resistance so that the current drawn is very low, of the order of 5×10^{-11} A. Since the resistance of the cell may be as large as 10^8 ohms, this means a voltage drop of 0.005 V:

$$E = I \times R = 5 \times 10^{-11} \times 10^8 = 0.005$$

or an error of 0.5% at 1.000 V.

The direct-reading instruments are the most popular today, particularly since they can be line-operated. Several commercial models are available from such companies as Beckman, Fisher, and Corning. The operating instructions are either printed directly on the instrument or are furnished in a pamphlet by the manufacturer.

EXPERIMENT 8.1. Acid-Base Titrations

The following experiments are chosen to illustrate the types of titration curves obtained with a strong acid, weak acid, and a polyprotic acid. The data can be used for standardizing a solution, analyzing an unknown, or determining the dissociation constant of a weak acid.

A typical experimental setup is shown in Fig. 8.2. The instructor will explain to you the operation of the pH meter and the precautions you should observe.

Procedure

(a) Strong Acid–Strong Base. Prepare solutions of about 0.1 N hydrochloric acid and sodium hydroxide as directed in Experiment 3.1, page 51. Then pipet a 25.00-ml aliquot of the acid into a 250-ml beaker and dilute the solution to about 100 ml with distilled water. Insert the electrodes into the solution, being sure that they dip about $\frac{1}{2}$ in. below the surface. Adjust the mechanical or magnetic stirrer

Fig. 8.2 Potentiometric titration using pH meter.

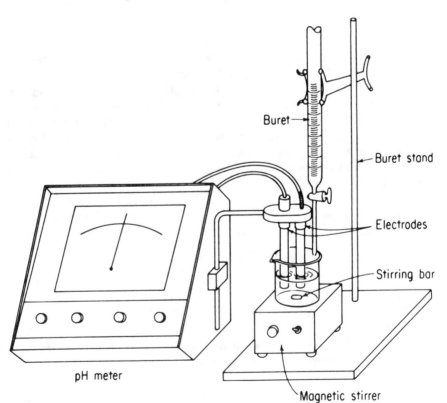

and set up a buret containing the base solution as shown in Fig. 8.2. Measure and record the *p*H of the solution before the addition of any titrant. Then add from the buret about 5 ml of the base solution and again measure the *p*H. Record this value as well as the buret reading at this point.

Proceed in this manner to record the *p*H and buret readings after the addition of about 10, 15, and 20 ml of titrant. Then add the titrant in about 1-ml intervals until the equivalence point is almost reached. (It may save time to run rapidly through the first titration to approximately locate the equivalence point. The second titration is then done carefully.) Then add the titrant in 0.1-ml intervals until the equivalence point is passed. It will be evident when this point is reached because of the large change in *p*H that occurs. Finally, record two additional readings at about 5 and 10 ml of excess titrant.

Make the following plots of the data: (1) *p*H vs. ml of NaOH; (2) $\Delta pH/\Delta V$ vs. ml of NaOH; and (3) $\Delta^2 pH/\Delta V^2$ vs. ml of NaOH. Determine the volume of base required by the acid from each plot and also calculate the volume by the analytical method described in Chapter 12 of the text. Calculate the volume ratio of the two solutions.

Repeat the titration of two additional aliquots of acid (Note), adding to the first 2 drops of methyl red indicator and to the second 2 drops of phenolphthalein. Note the *p*H values at which these indicators change color and compare these with the data given in Table 6.3 of the text. Average the values obtained for the volume ratios and calculate the precision of the measurement.

Note

If standard solutions are already at hand, one titration is sufficient to illustrate the titration curves.

(b) Weak Acid–Strong Base. Weigh a sample of about 0.7 to 0.9 g of pure, dry potassium acid phthalate (Note) on the analytical balance. Dissolve the sample in about 100 ml of distilled water and titrate with the sodium hydroxide. Measure the *p*H at different increments of titrant as in part (a).

Plot the titration data in the same manner as above and compare the curves obtained for the weak acid with those obtained for the strong acid. Justify the selection of the indicator used in Experiment 3.3, page 49. Calculate the normality of the base and acid solutions. If further use is to be made of these solutions the standardization should be repeated (consult the instructor).

Note

If the normalities of the solutions are already known, a sample of unknown purity can be titrated.

(c) pK of an Unknown Acid. Dissolve a sample of an unknown acid (consult the instructor) in about 100 ml of distilled water in a 250-ml beaker and titrate the solution with standard sodium hydroxide as described in part (a). Plot the data as *p*H vs. ml of NaOH and determine the equivalence volume. Read from the curve the *p*H at half the volume required to reach the equivalence point. At this halfway point *p*H $= pK_a$, and the acid constant is thus determined. Report this value to your instructor and repeat the titration if so directed. The instructor may also wish to know the concentration of the unknown acid, if this is a solution, or the equivalent weight of the unknown if it is a solid. If the sample is a solid, it should be weighed on an analytical balance before it is dissolved.

(d) Titration of Phosphoric Acid. Pipet 25.00 ml of a phosphoric acid solution of unknown concentration into a 250-ml beaker. Dilute the solution to about 100 ml, insert the electrodes, and titrate with standard sodium hydroxide as directed in part (a). You should observe two breaks in the titration curve, one around *p*H 4 to 5 and the other around *p*H 9 to 10. Plot the titration curve as *p*H vs. ml of NaOH. Determine the following from the curve and report to the instructor: (a) the molarity of the acid solution and (b) the values of pK_{a_1} and pK_{a_2} for the first two dissociation constants of phosphoric acid (Note).

Note

For acids whose dissociation constants are greater than about 10^{-3} to 10^{-2} an appreciable error is made by using the expression *p*H $= pK_a$ halfway to the equivalence point. The following expression can be used for such stronger acids:

$$pH = pK_a - \log \frac{C - [H^+]}{C + [H^+]}$$

The volume to the equivalence point, V_e, is first determined; then the *p*H at $\frac{1}{2} V_e$ is read and the value of $[H^+]$ calculated. The term C is the concentration of the anion halfway to the equivalence point and is given by

$$C = \frac{\frac{1}{2} V_e \times N}{\text{total volume}}$$

where the total volume is $\frac{1}{2} V_e$ + the starting volume. The latter volume should be known to ± 10 ml.

EXPERIMENT 8.2.   Redox Titrations

There are many redox titrations that can be used to illustrate the potentiometric technique. In this experiment the titration of iron(II) with dichromate or cerium(IV) solution is carried out. The indicator electrode is platinum and the reference is a saturated calomel electrode.

Procedure

(a) Titration of Iron with Dichromate. Prepare a standard solution of potassium dichromate by weighing accurately about 1.25 g of the pure, dry salt.

Dissolve the salt in water, transfer the solution to a 250-ml volumetric flask, and dilute the solution to the mark. Fill a buret with the solution.

Weigh accurately a sample of about 1.3 to 1.5 g of pure iron(II) ammonium sulfate. Dissolve the salt in about 100 ml of distilled water and add about 10 ml of concentrated sulfuric acid.

Insert a platinum and a saturated calomel electrode into the solution and adjust the stirrer if one is to be used. If a *p*H meter is to be employed, set the instrument to measure potential rather than *p*H. The platinum wire is the positive electrode. Add about 5 ml of the dichromate solution from the buret and then measure the potential. Continue the titration in the usual manner, recording the potential and volume of dichromate solution, until about 10 ml of excess titrant is added.

Make the following plots of the data: (1) potential vs. ml of titrant, (2) $\Delta E/\Delta V$ vs. ml of titrant, and (3) $\Delta^2 E/\Delta V^2$ vs. ml of titrant. Determine the volume of dichromate used, and from this volume and the weight of iron(II) ammonium sulfate, calculate the normality of the dichromate solution. Compare this value with the normality calculated from the weight of dichromate dissolved in 250 ml of solution. Repeat the titration if desired (consult the instructor).

(b) Titration of Iron with Cerium(IV). Prepare 250 ml of a standard solution from cerium(IV) ammonium nitrate as directed on page 74. Dissolve a sample of iron ore and reduce it with tin(II) chloride as directed in Experiment 4.5. Stop the procedure after adding excess tin(II) chloride; that is, do not remove the excess tin(II) ion with mercury(II) chloride. The titration curve will show two breaks: the tin reacting first, followed by the iron. The difference in volumes to the two end points gives the volume of titrant used by the iron sample.

Insert a platinum and a saturated calomel electrode into the iron solution in a 250-ml beaker. Adjust the volume to about 100 ml and set the *p*H meter to measure potential rather than *p*H. The platinum wire is the positive electrode. Add about 5 ml of the cerium(IV) solution from a buret and measure the potential. Continue the titration in the usual manner, recording the potential and volume of cerium(IV) solution, until about 10 ml of excess titrant is added.

Plot the potential vs. ml of titrant and measure the difference in volume between the first and second breaks. From this volume and the weight of the sample, calculate the percentage iron in the sample. Report this value to the instructor and repeat the determination if so instructed.

Also plot $\Delta E/\Delta V$ vs. ml of titrant (for the titration of iron) and $\Delta^2 E/\Delta V^2$ vs. ml of titrant and compare these with the plot obtained in Experiment 8.1(a).

EXPERIMENT 8.3. **Precipitation Titrations**

A piece of polished silver wire serves as the indicator electrode for titrations involving silver ions. This electrode is available commercially for use with *p*H meters. A calomel electrode cannot dip directly into the solution being titrated since silver chloride would precipitate; instead, this electrode is placed in a saturated potassium nitrate solution in a separate vessel, and the two solutions are joined by a salt bridge containing potassium nitrate.[1] Directions are given for the titration of chloride with silver nitrate, and for the titration of a chloride-iodide mixture.

Procedure

(a) **Titration of Chloride.** Prepare a standard solution of silver nitrate by weighing accurately about 4.5 g of the pure salt. Dissolve the salt in water, transfer the solution to a 250-ml volumetric flask, and dilute the solution to the mark.

Accurately weigh a dried chloride sample of unknown purity (consult instructor as to the weight required) and dissolve the salt in about 100 ml of distilled water. Insert the silver electrode and the potassium nitrate salt bridge into the solution and adjust the stirrer. Set the *p*H meter to read potential. The silver wire is the positive electrode. Add about 5 ml of the silver nitrate solution from the buret and then measure the potential. Continue the titration in the usual manner, recording the potential and volume of silver nitrate, until about 10 ml of excess titrant is added.

Make the following plots of the data: (1) potential vs. ml of silver nitrate, (2) $\Delta E/\Delta V$ vs. ml of silver nitrate, and (3) $\Delta^2 E/\Delta V^2$ vs. ml of silver nitrate. Determine the volume of titrant, and from this volume and the molarity calculate the percentage of chloride in the sample. Repeat the titration if desired (consult instructor).

(b) **Titration of a Chloride-Iodide Mixture.** The standard solution of silver nitrate is prepared as in part (a). Secure from the instructor a solution containing both chloride and iodide ions of unknown concentration. Pipet a 25-ml aliquot into a 250-ml beaker and dilute the solution to about 100 ml with water. Insert the silver electrode and potassium nitrate salt bridge and adjust the stirrer. Set the *p*H meter to read potential. Titrate with silver nitrate in the usual manner, recording the potential and volume of titrant, noting carefully the volumes near the two equivalence points.

Plot the potential vs. ml of titrant and determine the volumes of titrant used by iodide and chloride. Calculate the molarities of these two ions and report the results to the instructor.

EXPERIMENT 8.4.   Identification of an Amino Acid[2]

The titration of amino acids is discussed in some detail in Chapter 7 of the text. The object of this experiment is to identify an amino acid by potentiometric titration

---

[1] Since the *p*H of the solution changes little during the precipitation titration, it is possible to use a glass electrode in place of calomel as the reference electrode, thereby eliminating the salt bridge.

[2] This experiment was designed by Ronald C. Johnson of Emory University, who has kindly given us permission to use it here.

TABLE 8.1   SOME AMINO ACIDS

Name	Molecular Weight	pK_a
Alanine	89.1	9.87
Arginine · HCl	210.7	1.82, 9.09
Asparagine	150	8.80
Aspartic acid	133.1	3.86, 10.0
Cysteine · HCl	176.6	1.71, 8.33, 10.78
Cystine	240.3	8.02, 8.71
Glutamic acid	147	4.25, 9.67
Glutamic acid · HCl	183.6	2.19, 4.25, 9.67
Glycine	75.1	9.87
Histidine	155.2	6.02, 9.08
Histidine · HCl	209.6	1.7, 6.02, 9.08
Leucine	131.2	9.74
Lysine · 2HCl	219.1	2.20, 8.95, 10.53
Methionine	149.2	9.21
Phenylalanine	166.2	9.24
Proline	115.1	10.60
Taurine	125.2	8.74
Tyrosine	181.2	9.19, 10.47
Valine	117.2	9.72

with NaOH. From the data obtained, the molecular weight and pK_a value (or values) can be estimated. The number of possible unknowns will be limited to those shown in Table 8.1. All of the compounds have a weakly acidic $^+NH_3$ group and will give at least one end point similar to that shown in Figure 7.3 of the text. In order to get an appreciable $\Delta pH/\Delta V$, relatively concentrated solutions are titrated. It should be kept in mind that one can locate the end point for the titration of a very weak acid group from the end point of a stronger acid group, provided, of course, that such a group is present in the molecule.

Procedure

Secure from the instructor your amino acid unknown and record its number in your notebook. Do not dry the sample since it may decompose on heating. Weigh accurately a sample of about 0.3 g (Note 1) of the acid into a 100-ml beaker. Dissolve the sample in about 20 to 30 ml of water, using heat if necessary. (Cool the solution before titration.) The volume of water should be kept at a minimum but must be sufficient to cover the stirring bar so that the tips of the electrodes can be immersed without being hit by the stirring bar. Titrate with standard NaOH (about 0.1 N), using 1.0-ml increments throughout the titration. If the pH changes by more than 0.3 unit/ml of base added, 0.5-ml increments would be appropriate. Continue the titration until the pH reaches a value of about 11.8.

Prepare a graph plotting pH vs. ml of NaOH. From the graph determine whether your acid is mono-, di-, or triprotic, determine its pK_a value(s) (Note 2), and its molecular weight. Identify the amino acid on the basis of your data and the information given in Table 8.1. Since many of the acids do not give sharp breaks on titration and since the acids may not be completely pure, your molecular weight value may differ from the correct value by as much as 2%. The pK_a values may differ from those reported in Table 8.1 by 0.2 pK unit.

Turn in your graph to the instructor. Report the identity of your unknown and its molecular structure. (Consult a book on organic or biochemistry). Finally, identify the acid group from which a proton is being removed in each titration step.

Notes

1. To the instructor: Compounds 1 to 5, 7 to 9, 12, 13, 15, and 19 are available and have been titrated successfully by beginning students. You may wish to confine your unknowns to this group of acids.

2. For acids with pK_a values of 3 or less the exact expression given on page 149 should be used.

9

Electrolysis and Polarography

The principles of electrolysis and polarography are discussed in Chapter 13 of the text. Directions are given here for experiments in electrogravimetry, coulometry, polarography, and amperometric titrations.

EXPERIMENT 9.1. Electrolytic Determination of Copper

An experiment involving the deposition of a metal by electrolysis is normally included in the introductory course in quantitative analysis. The determination of copper is one of the classic illustrations of the electrogravimetric technique. Directions are given here for the determination of copper in a solution free of interfering metals.

Apparatus

A schematic diagram of the necessary apparatus is given in Fig. 13.2 of the text. The source of direct current is often a 6-V rectifier rather than a storage battery, and an ordinary rheostat of suitable size serves as the adjustable resistance. An inexpensive unit that can be easily assembled has found wide usage.[1] It is desirable to provide mechanical stirring if the equipment is available. Commercial "electroanalyzers" which provide for rotation of one electrode to effect stirring can be purchased. Such an apparatus is shown in Fig. 9.1.

[1] P. J. Elving, J. R. Hayes, and M. G. Mellon, *J. Chem. Educ.*, **30**, 254 (1953).

Fig. 9.1 Commercial apparatus for electrodeposition. (Courtesy of E. H. Sargent and Co.)

The electrodes are usually platinum, although copper gauze can be employed as the cathode for the deposition of copper. The cathode is usually a cylinder of platinum gauze, and the anode is a spiral of platinum wire or a small cylinder if the electrode is rotated.

Procedure

Clean the platinum electrodes by immersing them in warm $1:3$ nitric acid for about 5 min. Then rinse them well with tap water and distilled water. Place the gauze cathode on a watch glass and dry it in an oven at about $105°C$. Cool the cathode in a desiccator and then weigh it on the analytical balance. Avoid touching the gauze with the fingers as this may leave grease on the surface and prevent copper from adhering.

The solution to be electrolyzed should contain about 2 ml of sulfuric acid and 1 ml of nitric acid (Note 1) per 100 ml. Obtain this solution or a solid unknown from the instructor (Note 2). Connect the electrodes properly to the apparatus, placing the spiral anode inside the gauze cylinder. Make sure that the electrodes do not touch. Raise the beaker (a tall-form one is convenient) around the electrodes and adjust the height so that the lower edge of the cathode almost touches the bottom of the beaker. About $\frac{1}{4}$ in. of the top of the cathode should not be covered by the solution. The solution can be diluted if necessary with distilled water. Cover the beaker with a split watch glass and adjust the rheostat to give its full resistance.

Turn on the stirrer and close the circuit to start the electrolysis. Only a small current should flow since the resistance is high. Gradually lower the resistance until the current is 2 to 4 A and the voltage is below 4 V (Note 3). Electrolyze at this voltage until the blue color of copper has disappeared (usually about 30 to 45 min). Add 0.5 g of urea (Note 4), continue the electrolysis for 15 min longer, and then add sufficient water to cover completely the top of the cathode (Note 5). Continue the electrolysis for an additional 15 min using 0.5-A current, and if no copper is deposited on the fresh cathode surface, the deposition is complete.

To stop the electrolysis, turn off the stirrer but do not interrupt the current at this time. Remove the support under the beaker, and slowly lower the beaker with one hand while washing the exposed portion of the cathode with a stream of water from the wash bottle. As soon as the cathode is completely out of the solution, cut off the current and raise a beaker of distilled water to cover the electrodes. Wash the electrodes with a second portion of distilled water and then disconnect the cathode. Dip the cathode into a beaker of acetone or alcohol and place it on a watch glass in the oven at about 105°C for 5 min (Note 6). Cool the electrode to room temperature and then weigh it accurately.

If a prepared solution was used, report the total weight of copper in your solution. If copper ore was employed, calculate the percentage of the metal in the sample. Duplicate samples should be run; consult the instructor as to his wishes.

The copper can be removed from the electrode by placing it in warm 1:3 nitric acid for a few minutes. The electrode is then rinsed well with tap water and distilled water.

Notes

1. If the acid has a yellow tinge indicating the presence of oxides of nitrogen, boil the solution until the oxides are expelled. Nitric acid improves the nature of the copper deposit by preventing the evolution of hydrogen at the cathode.

2. This solution can be prepared by dissolving pure copper foil in nitric acid according to procedure (b) of Experiment 4.15. It is then diluted in a volumetric flask and aliquot portions are given to the students. Alternatively, commercial unknowns of copper oxide which are readily dissolved in sulfuric acid can be used. If a brass sample is being analyzed, the filtrate from the lead determination by the sulfate method can be electrolyzed, after first adjusting the volume to about 100 ml and adding 1 ml of concentrated nitric acid.

3. If a mechanical or magnetic stirrer is not available, the electrolysis can be carried out without stirring by using a current of about 0.5 A. The electrolysis should then be allowed to run overnight.

4. Nitrite prevents complete deposition of copper and is removed by urea according to the equation

$$2NO_2^- + 2H^+ + (NH_2)_2CO \rightleftharpoons CO_2 + 2N_2 + 3H_2O$$

The nitrite is formed by the reaction

$$2H^+ + NO_3^- + 2e \rightleftharpoons H_2O + NO_2^-$$

5. If the solution is not to be used for further analysis, a few drops can be removed with a pipet and tested with concentrated ammonia. The deep blue color of the copper-ammonia complex indicates the incomplete deposition of copper.

6. Do not heat the electrode any longer than this since the surface of the copper becomes oxidized easily.

EXPERIMENT 9.2. **Separation of Copper and Nickel by Electrolysis**

Copper and nickel are on opposite sides of hydrogen in the activity series and hence can be separated electrolytically by control of *p*H. Copper is deposited from an acid solution and the electrode weighed. The hydrogen ion concentration is lowered by addition of ammonia and then nickel can be deposited. The separation is useful in the determination of these two metals in coinage alloys of copper and nickel and in Monel metal. These alloys usually contain a small amount of iron which must be separated before the electrolysis of nickel. Iron can be separated from nickel by precipitation of hydrous iron(III) oxide.

Procedure[2]

If the alloy (Note 1) is greasy, wash it with dilute ammonia, then with water, and finally with acetone. Dry the sample and weigh about 0.5 g into a 200-ml tall-form beaker. Add to the beaker 10 ml of water, 1 ml of concentrated sulfuric acid, and 2 ml of concentrated nitric acid. When the sample has dissolved, boil the solution to 100 ml and electrolyze the copper as directed in Experiment 9.1.

Wash the copper deposit thoroughly, taking care to save all the washings. Dry and weigh the electrode and return it to the apparatus without removing the copper (Notes 2 and 3).

Evaporate the filtrate slowly and carefully on a hot plate until fumes of sulfur trioxide appear (Note 4). Cool the residue and add to it carefully about 25 ml of water. Add about 10 ml of 1:1 ammonia to precipitate the iron. Filter off the hydrous oxide using a small filter paper and catch the filtrate in an electrolytic beaker. Wash the precipitate three times with small portions of water and then set the beaker containing the filtrate aside. Dissolve the precipitate by pouring over the filter paper a few milliliters of hot 3 *M* sulfuric acid, catching the filtrate in the original electrolytic beaker. Wash the paper with water and again precipitate the iron by adding 10 ml of 6 *M* ammonia. Filter the precipitate through the same paper and wash it with water, receiving the solution in the original beaker which contains the nickel solution.

[2]H. H. Willard, N. H. Furman, and C. E. Bricker, *Elements of Quantitative Analysis*, 4th ed., D. Van Nostrand Co., Inc., New York, 1956, p. 438.

The hydrous oxide can be ignited to Fe_2O_3 and weighed if desired (consult instructor). Otherwise, discard the precipitate.

To the filtrate containing the nickel, add 15 ml of concentrated ammonia and dilute the solution to about 100 ml. Electrolyze the nickel in the same manner as directed for copper. The completeness of deposition can be judged by the disappearance of the blue color of the nickel-ammonia complex. A few drops of the solution can be removed and tested for nickel with dimethylglyoxime.

When the deposition of nickel is complete, remove the solution and wash the electrode in the same manner as directed for copper. Rinse the electrode twice with distilled water and once with acetone or alcohol. Dry the electrode for 5 min in the oven, then cool and weigh it in the usual manner. Report the percentages of copper and nickel in the alloy.

Notes

1. The instructor may prefer to use prepared solutions of copper and nickel rather than the alloy. An aliquot portion is given each student, who reports the weight of each metal in the solution. About 2 ml of concentrated sulfuric acid is added, the solution is diluted to 100 ml, and copper is electrolyzed. No nitric acid is added, since it interferes with the deposition of nickel. The lengthy procedure required to remove iron is avoided if such solutions are used.

2. Nickel can be deposited on top of the copper.

3. If a copper-nickel solution is being analyzed, neutralize the solution at this point with concentrated ammonia (litmus), add an additional 15 ml of concentrated ammonia, and electrolyze the nickel as directed above.

4. This operation removes nitrates and oxidizes iron to the $+3$ state.

EXPERIMENT 9.3.   Separation of Copper, Bismuth, and Lead by Controlled Potential Electrolysis

The method of controlled potential electrolysis for separating metals which have standard potentials close together has been discussed in Chapter 13. The present experiment gives directions for separating copper, bismuth, and lead by regulation of the cathode potential. It is based on the method described by Lingane and Jones,[3] in which a tartrate solution of controlled pH is used. A potentiostat is needed for the experiment.

Procedure

Secure from the instructor a solution containing copper, bismuth, and lead of unknown concentrations (Note 1). Pipet a 50-ml aliquot into a 250-ml beaker and dilute with water to about 150 ml. Add 1.5 g of urea, 12 g of sodium tartrate

[3] J. J. Lingane and S. L. Jones, *Anal. Chem.*, **23**, 1798 (1951); see also J. J. Lingane, *Electroanalytical Chemistry*, 2nd ed., Interscience Publishers, Inc., New York, 1958, p. 397.

dihydrate, and 1.0 g of succinic acid (Note 2). Stir the solution, add 2 g of hydrazinium chloride, and dilute to about 200 ml with water. Adjust the pH to 5.9 (pH meter) by slowly adding 2M NaOH. Be sure the solution is below 30°C before starting the electrolysis.

Platinum electrodes are to be employed. They are cleaned by immersing them in warm 1 : 3 nitric acid for about 5 min. Then rinse them well with tap water and distilled water. Place the gauze cathode on a watch glass and dry it in an oven at about 105°C. Cool the cathode in a desiccator and then weigh it on the analytical balance. Connect the electrodes to the potentiostat; a saturated calomel electrode serves as the reference. Adjust the beaker so that about $\frac{1}{4}$ inch of the top of the cathode is not covered by the solution.

Turn on the stirrer and electrolyze the copper at −0.30 V vs. SCE until the blue has disappeared. Add sufficient water to cover the top of the cathode and continue the electrolysis for a few minutes to see if the copper is completely deposited. Turn off the stirrer but do not interrupt the current at this time. Remove the support under the beaker, and slowly lower the beaker with one hand while washing the exposed portion of the cathode with a stream of water from the wash bottle. As soon as the cathode is completely out of the solution, cut off the current and rinse the cathode with either acetone or alcohol. Place the cathode on a watch glass and dry it in an oven at 105°C for about 5 min. Cool the electrode to room temperature and then weigh it accurately.

Connect the cathode to the potentiostat without removing the copper deposit (Note 3). Electrolyze the bismuth at −0.40 V vs. SCE, noting that the current rises slowly after the deposition begins. After about 45 min, discontinue the electrolysis as above, rinse, dry, and weigh the electrode.

Connect the cathode to the potentiostat again without removing the copper and bismuth. Electrolyze the lead at −0.60 V vs. SCE. After about 40 min, test the solution for lead by adding a drop of it to a drop of potassium dichromate. If no yellow precipitate of lead chromate is observed, the electrolysis can be discontinued. Rinse, dry, and weigh the electrode (Note 4).

Report to the instructor the number of milligrams each of copper, bismuth, and lead in the solution.

Notes

1. The concentrations of copper and bismuth should be in the range 0.5 to 2.0 mg/ml and that of lead from 0.2 to 0.5 mg/ml in the final solution. The final volume will be about 200 ml.

2. The tartrate concentration should be between 0.25 and 0.50 *M*. If more than 100 mg of copper is present (consult instructor), the amount of succinic acid should be increased: 1 g/100 mg of copper. The succinic acid-succinate system buffers the solution and prevents the pH from dropping below 5.2 during the electrolysis of copper.

3. Bismuth and lead should not be deposited directly on a platinum surface as the platinum may be destructively alloyed.

4. The metals can be removed by dipping the electrode into concentrated nitric acid.

EXPERIMENT 9.4. Coulometric Titration of Arsenite with Iodine

Coulometric titrations are discussed in detail in Chapter 13 of the text. Directions are given here for the titration of arsenite ion with electrically generated iodine. The iodine is produced by the oxidation of iodide ion at a platinum electrode using a constant-current source:

$$2I^- \; \rightleftharpoons \; I_2 + 2e$$

The cathode is platinum also, and the reaction at this electrode is the reduction of hydrogen ion:

$$2H^+ + 2e \; \rightleftharpoons \; H_2$$

The number of coulombs passed is calculated from the value of the constant current and the elapsed time. From the number of coulombs, the equivalents of iodine generated, and consequently the equivalents of arsenic in the sample, can be calculated from Faraday's law. The indicator is starch, which gives the familiar deep blue color with excess iodine.

Some source of constant current is required. Several commercial coulometric power supplies are available, or an apparatus may be assembled in the laboratory. Meloan and Kiser[4] describe a simple, inexpensive circuit which is said to be very satisfactory. Another is described by Reilley and Sawyer.[5]

Procedure

Secure a sample of arsenic(III) oxide of unknown purity and dry it at 105°C for 1 h. Weigh accurately a portion of about 0.25 to 0.30 g (Note) into a 100-ml beaker, add 10 ml of water in which about 1 g of NaOH has been dissolved, and stir until the solid has dissolved. Add about 40 ml of water and then 1 : 1 HCl until the solution is slighty acidic. Transfer the solution to a 500-ml volumetric flask and dilute to the mark.

Run a blank determination as follows: Add to a 150-ml beaker about 50 ml of water, 0.8 g of pure KI, about 2 g of reagent-grade $NaHCO_3$, and 5 ml of starch solution (see Note 6, Experiment 4.12, for directions). Insert two platinum foil electrodes in the solution, adjust the instrument to give a current of about 10 mA, start the stirrer, and simultaneously start the electrolysis current and the timer. Stop the titration as soon as the deep blue color persists in the solution for at least 30 s.

Record the time and reset the timer. Using a 25-ml pipet, add an aliquot of the arsenic solution to the solution in the beaker. Start the stirrer and start simul-

[4]C. E. Meloan and R. W. Kiser, *Problems and Experiments in Instrumental Analysis*, Charles E. Merrill Publishing Company, Columbus, Ohio, 1963, p. 171.

[5]C. N. Reilley and D. T. Sawyer, *Experiments for Instrumental Methods*, McGraw-Hill Book Company, New York, 1961, p. 366.

taneously the electrolysis current and the timer. Stop the titration as soon as the deep blue color persists in the solution for at least 30 s.

Repeat the entire procedure on two additional samples. Calculate the percentage of As_2O_3 in the sample.

Note

This weight of sample is calculated to require 4 to 5 min for the titration assuming that the sample is about 10% As_2O_3 and that the current is about 10 mA. Consult the instructor for any necessary changes.

EXPERIMENT 9.5.   Polarographic Determination of Lead

A wide variety of polarographs, both manual and recording, are available commercially. An inexpensive apparatus that can be readily assembled in the laboratory has been described by Lingane.[6] A complete setup is now available from the Heath Co. based on its polarography module.

The commercial houses that sell polarographs also supply dropping-mercury electrode assemblies, including capillaries of marine barometer tubing already tested for drop time,[7] and a wide variety of cells. One convenient arrangement of such apparatus is shown in Fig. 9.2. The cell shown has a pool of mercury on the bottom for the anode and is easily cleaned between runs. The H-type cell shown in Figure 13.8 of the text provides a very convenient arrangement for a permanently attached reference electrode.

This experiment is designed to illustrate conventional polarography with the dropping-mercury electrode and its application to analysis.

Preliminary Instructions

All parts of the mercury electrode assembly should be thoroughly cleaned with 1:1 nitric acid, rinsed well with distilled water, and finally with acetone. Allow the parts to dry thoroughly before adding mercury. For making connections, use Tygon tubing which has been thoroughly rinsed with water and then dried. Draw some of the nitric acid up through the capillary by applying suction to one end. Then draw distilled water and finally acetone through the capillary and leave the suction on until the electrode is dry.

Triply distilled mercury is available commercially and can be used without further purification. If this is not available, the mercury should be cleaned by

---

[6] J. J. Lingane, *Anal. Chem.*, **21**, 47 (1949).

[7] This tubing can be bought directly from the Corning Glass Works, Corning, N.Y., and cut to suitable lengths.

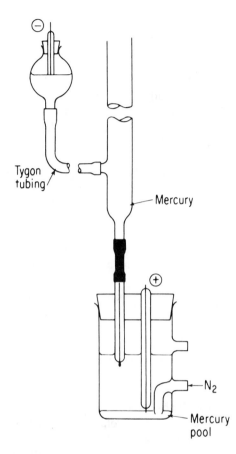

Fig. 9.2 Dropping-mercury electrode assembly and cell.

dropping it through a glass tube about 1 m high containing 3% nitric acid. It is then distilled under reduced pressure (Note).

Add sufficient mercury to the reservoir and start the flow through the capillary by applying suction. Immerse the end of the capillary in distilled water and adjust the height of the mercury until the drop time is between 4 and 6 s/drop. Always leave the tip of the electrode immersed in distilled water when it is not in use. If the end becomes plugged, insert it in concentrated nitric acid for a few moments and then in distilled water.

Note

Extreme care should be used in handling mercury so as not to spill the liquid. It is convenient to place a large tray containing some water under the mercury electrode assembly. If any mercury is spilled outside the tray, it should be cleaned up immediately. Mercury vapor is poisonous. Gold rings and watches should not be worn while handling mercury.

Procedure

A. Preparation of Stock Solutions

(a) *1 M KCl.* Dissolve 18.64 g of reagent-grade KCl in water, transfer to a 250-ml volumetric flask, and dilute the solution to the mark.

(b) *0.01 M PbCl$_2$.* Weigh accurately about 0.69 g of PbCl$_2$ (0.6955 g for exactly 0.01 M), dissolve the salt in water, and transfer the solution to a 250-ml volumetric flask. Dilute the solution to the mark.

(c) *0.2% Triton X-100.* Dissolve 0.2 g of the reagent in 100 ml of water.

(d) *Unknown lead solution.* This solution, furnished by the instructor, is approximately 0.01 M.

B. Preparation of Test Solutions

1. *Blank solution.* To a 100-ml volumetric flask add 10 ml of 1 M KCl and 1.0 ml of Triton X-100. Dilute the solution to the mark with distilled water.

2. *Solutions for wave height-concentration plot.* Prepare five solutions of known lead concentration in 100-ml volumetric flasks as follows: Place 10 ml of 1 M KCl and 1.0 ml of Triton X-100 in each flask. Then add the following quantities of the standard lead solution: 2.00 ml to the first flask, 5.00 ml to the second, 10.00 ml to the third, 15.00 ml to the fourth, and 20.00 ml to the fifth. Dilute the solution in each flask to the mark with distilled water.

3. *Solution with no maximum supressor.* Prepare another solution by placing 10 ml of 1 M KCl and 20.00 ml of the lead solution in a 100-ml volumetric flask. Dilute the solution to the mark.

4. *Unknown solution.* Add 10.00 ml of the unknown lead solution, 10 ml of 1 M KCl, and 1.0 ml of Triton X-100 to a 100-ml volumetric flask. Dilute the solution to the mark with distilled water.

Run the polarograms as follows. Transfer a portion of the solution to be electrolyzed to the cell after first rinsing the cell with the solution. Purge the solution with prepurified nitrogen for a few minutes to remove dissolved oxygen. Run nitrogen over the solution while the polarogram is recorded. Record the polarogram as directed by the instructor. Secure polarograms with the following test solutions prepared under Section B above: (1) blank, (2) the five solutions of known concentrations, (3) the solution with no maximum suppressor, and (4) the unknown solution.

Determine the value of $m^{2/3}t^{1/6}$ as follows: Replace the electrolysis cell with a small weighing bottle containing some 0.1 M KCl solution. Collect about 20 drops of mercury, measuring the time interval with a stopwatch. No voltage is applied to the cell. Remove the KCl solution with a medicine dropper, rinse the mercury with distilled water, then with acetone, and dry it for a few minutes in air. Weigh the bottle plus mercury and then weigh the empty bottle. Divide the time in seconds by the number of drops to get the drop time t. Divide the milligrams of mercury by the time in seconds to get the value of m in mg/sec.

Make the following calculations:

1. Measure the height of each wave obtained with the standard solutions in microamperes. Correct each for the residual current using the blank or extrapolation. Plot the wave height against the concentration, and from the wave height of the unknown, calculate the concentration of lead in the unknown solution.

2. From the wave height of the unknown and the value of $m^{2/3}t^{1/6}$, calculate the concentration of the unknown solution from the Ilkovic equation. The diffusion current constant, I, of lead is 3.80. The diffusion current must be in microamperes and the concentration units are millimoles per liter.

3. From any one of your polarograms, make a plot of the applied potential E vs. $\log i/(i_d - 1)$. The iR drop through the solution can be assumed to be negligible. Draw the best straight line through the points and calculate the value of n from the slope. Also read off the value of $E_{1/2}$ from the point on the line where the log term is zero, and compare this with the value obtained directly from the polarogram.

Report your results as desired by the instructor.

EXPERIMENT 9.6.   Amperometric Titration of Lead with Dichromate

The end point of the titration of lead with dichromate can be detected conveniently with the amperometric technique.[8] Lead chromate is precipitated:

$$2Pb^{2+} + Cr_2O_7^{2-} + H_2O \rightleftharpoons 2PbCrO_4 + 2H^+$$

At -1.00 V, both lead and dichromate ions are reduced. Hence the titration curve passes through a minimum, as shown in Fig. 13.15(c) of the text. The titration can also be carried out at zero volts. Only dichromate is reduced at this potential.

Procedure

Preparation of Solutions

(a) *0.01 M Pb(NO₃)₂*. Weigh accurately about 0.33 g (0.3312 g for exactly 0.01 M) of $Pb(NO_3)_2$. Dissolve the salt in water, transfer the solution to a 100-ml volumetric flask, and dilute to the mark.

---

[8] I. M. Kolthoff and Y. D. Pan, *J. Am. Chem. Soc.*, **61**, 3402 (1939).

(b) *0.20 M KNO₃ + buffer.* Dissolve 2.0 g of KNO_3, 2.0 ml of glacial acetic acid, and 0.82 g of sodium acetate in water and dilute to 100 ml.

(c) *0.05 M K₂Cr₂O₇.* Weigh accurately about 1.47 g (1.4710 g for exactly 0.05 *M*) of reagent-grade potassium dichromate. Dissolve the salt in water, transfer the solution to a 100-ml volumetric flask, and dilute to the mark.

(d) *0.2% Gelatin.* Dissolve 0.2 g of gelatin powder in 100 ml of hot water.

(e) *Unknown Pb(NO₃)₂ solution.* This solution furnished by the instructor is approximately 0.01 *M*.

Pipet 5.00 ml of the standard lead nitrate, 10 ml of the KNO_3–buffer solution, and 1 ml of gelatin into an electrolysis cell. Bubble nitrogen through the solution for 10 min to remove dissolved oxygen. Insert the dropping-mercury electrode and adjust the flow of mercury to about 5 s/drop. Set the applied potential at -1.00 V and set the current sensitivity to get as near full-scale deflection as possible.

Add about 0.1 ml of the standard potassium dichromate solution from a microburet. Bubble nitrogen through the solution for 1 min to mix the solutions and remove oxygen. Then record the current and the volume of titrant. Repeat this procedure until about four points are obtained before and after the equivalence point (about 0.5 ml). The current decreases at first and then increases after excess dichromate has been added. Do not exceed the equivalence point by more than about 0.2 ml.

Titrate in the same manner as above 5.00 ml of the unknown lead solution. Then plot on linear graph paper the current (scale divisions are satisfactory units) against the volume of titrant for the two titrations. Draw straight lines through the points before and after the equivalence point until they intersect. Read the volume of dichromate at the intersection. Corrections for dilution may be made as explained in Chapter 13 of the text. Compare the volume of titrant used by the first solution with that calculated from the known concentrations. Calculate the molarity of the unknown and report this to the instructor. He may ask you to do another titration.

Set the voltage at 0.00 V and repeat the titration of the lead solution of known concentrations. The removal of oxygen can be omitted. Note the shape of the titration curve and compare it to the one obtained at -1.00 V.

10

Chromatography and Ion Exchange

The principles of chromatography and ion exchange are discussed in Chapters 17 and 18 of the text. Since there are a large number of gas and liquid chromatographs on the market, only very general directions for the suggested experiments can be given here. Students should familiarize themselves with the instructions for operating the particular equipment provided and heed carefully the advice of the instructor.

GAS-LIQUID CHROMATOGRAPHY

There are three particular problems with nearly any conventional GLC setup: (1) the cylinder of compressed carrier gas is potentially dangerous; it should be anchored firmly, opened properly with the reducing valve closed, and shut down correctly; (2) the column should be kept below a temperature determined by the stability and volatility of the stationary liquid phase; and (3) the conventional thermal conductivity detector must be "off" unless carrier gas is flowing through the system. Directions provided in the laboratory should be followed carefully. It is recommended that the student have time to familiarize himself with the apparatus and to practice injecting samples before performing the experiments. Constancy of injection technique is very important in GLC analysis. Microliter syringes are expensive, easily damaged, and potentially dangerous; do not lay them down where they may be knocked to the floor, do not run around with them, and do not do anything that might snap the needles or dull or bend their tips.

EXPERIMENT 10.1   Behavior of a Homologous Series
of Normal Alkanes, and
Analysis of a Hydrocarbon Mixture

Procedure

The samples and conditions mentioned here will normally yield good results, but the student should follow whatever directions and suggestions are provided by the instructor. The stationary phase on the column may be silicone oil, Nujol, tricresyl phosphate, squalane, dinonyl phthalate, or various other of the relatively less polar substrates. A column temperature of 100°C or somewhat less is satisfactory. Sample sizes and signal attenuation settings may be adjusted to obtain the desired peaks on the recorder. Generally, 2- to 5-μliter samples may be used. The carrier gas flow rate may be perhaps 50 ml/min. The column should be in temperature equilibrium with the oven, helium flowing through the system and the detector "on" and stable; the recorder should draw a fairly straight horizontal line.

Using a microliter syringe and injecting samples of appropriate size, obtain chromatograms of individual samples of pure *n*-pentane, *n*-hexane, *n*-heptane, *n*-octane, and *n*-nonane. In many cases, there will be a principal peak plus one or two small impurity peaks. Note the tiny "blip" on the recorder baseline when the sample is injected. This is due to the small pressure change in the system when the plunger is pushed in the syringe and it serves as a marker of the injection time. Also notice the small air peak shortly after the injection.

Also run a chromatogram on a mixture of the five hydrocarbons. Measure the retention time for each peak and compare it with the values for the hydrocarbons injected separately. Could you identify the components if the mixture were an "unknown"?

Plot the logarithm of the retention time vs. the number of carbon atoms for the hydrocarbon series and draw the best straight line through the points. You may either look up the logs or plot the T_r values themselves on semilog graph paper. Could you identify a normal alkane in an unknown sample if a sample of the pure compound was not at hand for direct comparison of T_r values?

Obtain a chromatogram of an unknown sample furnished by the instructor. Samples of petroleum ether or ligroin, various lighter fluids, nonleaded gasolines, or synthetic mixtures of pure hydrocarbon are possibilities. In the case of the gasoline, additional standards such as the branched-chain 2,2,4-trimethylpentane (isooctane) will be needed for complete identification of all components.

Accurate quantitative analysis by GLC requires careful calibration techniques which may be too time consuming for an introductory experiment. However, a rough, semiquantitative analysis is fairly easy. For an unknown sample (as directed by the instructor) measure the total area under all the peaks. For each component, divide the peak area by the total area and multiply the quotient by 100. This gives approximately the weight percent of that component in the mixture, on the assumptions that all the components are structurally similar and have about the same thermal conductivity values. The areas may be measured in various ways; consult the instructor.

EXPERIMENT 10.2.   Some GLC Parameters

This experiment may be done separately or in combination with Experiment 10.1 as the instructor directs.

Procedure

Obtain a chromatogram on a mixture such as the series of *n*-alkanes above. Measure retention times and baseline widths by extending tangents through the inflection points of the peaks to the baseline. Calculate *n*, the number of theoretical plates, using each peak. Divide *n* into the length of the column to obtain HETP. How independent is HETP of the particular component used to measure it?

Inject varying volumes (say, from 1 to 30 μliters) of a pure hydrocarbon, measure HETP for each case, and plot HETP vs. sample size. Does the sample size affect the value? Compare HETP extrapolated to zero sample size with the values obtained with finite quantities. What is the effect of sample size upon column performance?

Vary the carrier gas flow rate between about 10 and 120 ml/min, obtaining a chromatogram of a pure hydrocarbon at each setting. Calculate HETP for each case and plot this vs. the flow rate. Is the graph as predicted from the van Deemter equation? *Caution:* Be sure the flow rate does not go to zero, as the detector could be damaged. Allow time for stabilization after each change in flow rate. A simple soap bubble flow meter made from an old buret is convenient and accurate.

Keeping within recommended limits and allowing ample time for temperature equilibration, study the chromatogram of the hydrocarbon mixture at several temperatures, say, 60, 80, 100, and 120°C. What is the effect of temperature upon retention times (and the time required for the analysis), resolution of the individual componets, and HETP values?

EXPERIMENT 10.3.   GLC Analysis of Benzene in Cyclohexane[1]

This experiment involves the determination of benzene in cyclohexane using gas-liquid chromatography. This topic is discussed in detail in Chapter 17.

Procedure

Secure a stock solution of 10.0 v/v% benzene in cyclohexane from the instructor. With this solution prepare six standards in the range 0.2 to 1.0%. Use

[1] This experiment was designed by Hubert L. Youmans of Western Carolina University, who has kindly given us permission to use it here.

a 10-ml buret with a Teflon stopcock to measure the stock solution into 25-ml volumetric flasks. Dilute to volume with spectrographic-grade cyclohexane. The standards used in Experiments 7.10 and 7.11 can be used if available. Take a 25-ml volumetric flask to the instructor for an unknown sample and dilute it to volume with cyclohexane.

The column you will use is 10 ft long and 5 mm i.d. It is packed with 20% Carbowax 20M on 80- to 100-mesh Chromosorb W. Carbowax 20M is poly(ethylene glycol) of average molecular weight 20,000. Chromosorb W is made by calcining diatomite with sodium carbonate at about 900°C. The fluxing agent causes fusion of the smaller particles to form larger microamorphous silica aggregates, some of which contain cristobalite crystals.

To fill a syringe, first pump the plunger with the needle tip below the surface of the liquid. When all air is expelled, draw more than the required amount of liquid into the syringe. Then with the needle pointing upward adjust the plunger to the desired mark.

Insert the needle into the septum to the limit; keep the needle straight while inserting. Do not bend it. Push the plunger in, then withdraw the needle. The three steps should be a smooth, continuous action.

To obtain retention times, stop the chart, mark it at the pen, then inject the sample and start the chart simultaneously. The retention time, t_R, is the distance from the starting point to the maximum of a peak multiplied by the chart speed. In calculating n, the number of theoretical plates, chart distances may by used because chart speed is a constant that occurs in both numerator and denominator.

To obtain measured volumetric flow rates, F_{meas}, measure the time in seconds for a soap film in the bubbleometer to go from 0 to 10 ml. Divide the number of seconds into 600. This gives F_{meas} in ml/min.

Operate the column at 90°C with a helium flow rate of 60 ml/min. Run chromatograms of the standards and unknown. Plot peak area vs. concentration of benzene for the standards. From this calibration curve, determine the concentration of benzene in your unknown.

LIQUID CHROMATOGRAPHY

EXPERIMENT 10.4.   High-Performance Liquid Chromatography[2]

The instructor will give you a description of your chromatograph and directions for its use. Such instruments generally consist of a liquid reservoir and prefilter, pump, column, detector, and recorder. In this experiment the column and guard column are packed for reverse-phase HPLC. The stationary phase is nonpolar and the mobile phase is polar. A Whatman PXS-1025 column gives satisfactory performance. It has 10-μm silica gel as the solid support; hydrogens of $-OH$ groups

---

[2] This experiment was designed by Hubert L. Youmans of Western Carolina University, who has kindly given us permission to use it here.

have been replaced by octadecyl groups until the silica is 5% carbon by weight. Solutes distribute during elution between the octadecyl groups and the mobile phase. The column is 25 cm long and has an internal diameter of 4.6 mm. It is packed commercially by a ramjet technique to obtain bed uniformity, high density, and good stability. The guard column is packed with the same material as the PXS-1025 column. It does not have the efficiency per unit length of the main column; its purpose is to prolong the life of the main column, although some separation is effected by it.

This experiment involves the separation and determination of phthalic acid and naphthalene. Phthalic acid is produced commercially by the oxidation of naphthalene:

Gas chromatography can be used for the separation but the acid must first be converted to a volatile ester. HPLC provides a simpler and quicker separation procedure.

Procedure

Weigh about 1 g of the unknown into a 50-ml volumetric flask. First weigh the clean, dry flask to four decimal places on the analytical balance. Than add the unknown and weigh the flask again. The difference between the weighings is the sample weight.

Similarly prepare a standard solution by weighing about $\frac{1}{2}$ g of phthalic acid and $\frac{1}{2}$ g of naphthalene in a second flask. To each flask add methanol almost to the neck of the flask. Place the flasks in a boiling-water bath and swirl them gently until the solids dissolve. Cool and dilute to volume with methanol and mix thoroughly.

Using a pipet, transfer 10 ml of the unknown to each of four 50-ml volumetric flasks. Pipet 5 ml of the standard solution to one of the flasks, 10 ml to a second, and 15 ml to a third. The fourth flask contains only the unknown. Dilute the four solutions to volume with methanol and mix.

Run HPLC chromatograms of the four solutions. Chromatograph 10 μl of sample at an inlet pressure of about 625 psi. The eluent is 75:25 methanol–water; the flow rate should be constant for the four runs. Use a chart speed of 1 in./min; use the chart drive only when samples are in the chromatograph.

Standard Addition Plot. Besides presenting the chromatogram that is a record of the electrical output of the detector, the recorder electrically measures and records the area under the chromatographic curve. This integrated signal is obtained as electrical counts. The integrator pen moves horizontally at a rate proportional to the amount of solute represented by a chromatographic peak. The number of

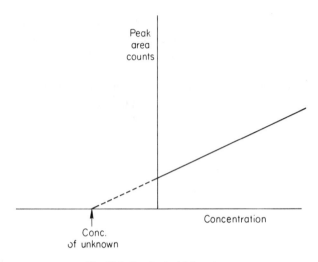

Fig. 10.1 Standard addition plot.

counts produced by the standards can serve as the signal variable of a calibration curve.

Make standard addition plots as follows. Plot peak area counts on the ordinate vs. concentration of the standards as the abscissa (Fig. 10.1). The value for the unknown is plotted as though its concentration is zero. Then draw a straight line through the points, extrapolating it to the concentration axis as shown in the figure. The left side of the abscissa is the mirror image of the right and the value of the intercept is the concentration of the unknown in each of the four flasks. From these data calculate the percent of each phthalic acid and naphthalene in the unknown.

This is a matrix elimination method. If the peak area counts fall on a straight line, effects of the matrix (i.e., the solvent and other solutes) are eliminated. The matrix gives results more reliable than does a direct method that depends upon a calibration curve. In cases where only one sample is to be analyzed, standard addition gives better results for less work than does a direct method.

EXPERIMENT 10.5.   **Ion-Exchange Separation and Spectrophotometric**
Determination of Nickel and Cobalt

The ion-exchange method of separation is discussed in Chapter 18 of the text. The present experiment illustrates this method by the separation of nickel and cobalt with a strong-base anion exchanger. In solutions containing a large concentration of hydrochloric acid, many metals are converted into complex anions (chlorides) and are adsorbed by an anion resin.[3] Other metals that are not readily converted

[3] See K. A. Kraus and F. Nelson, *Proceedings of the International Conference at Geneva*, Vol. 7, Session 9B.1, United Nations, 1956, p. 837.

into complex anions are not adsorbed by such a resin. Cobalt forms the deep blue complex, $CoCl_4^{2-}$, and this is strongly adsorbed form 9 M hydrochloric acid. Nickel is not adsorbed from such a solution, and a separation of nickel from cobalt is easily effected. The cobalt is readily removed by washing the resin with hydrochloric acid less concentrated than about 4 M.

The two metals can be determined by standard spectrophotometric procedures. Nickel is oxidized with bromine in ammoniacal solution and then treated with dimethylglyoxime. A wine-red or brown complex is formed, the identity of which has not been established. Cobalt is determined by conversion of the metal to the blue complex cobalt thiocyanate, $CO(CNS)_4^{2-}$. An acetone solution is used since the complex is somewhat dissociated in water unless a large excess of thiocyanate is added. Both procedures are based on those given by Sandell.[4]

Procedure

A. Prepare the Following Solutions:

(a) *Standard nickel solution.* Dissolve 0.405 g of uneffloresced crystals of $NiCl_2 \cdot 6H_2O$ in 100 ml of 0.1 M hydrocloric acid in a volumetric flask. Pipet 10 ml of this solution into another 100-ml volumetric and dilute to the mark with 0.1 M hydrochloric acid. This second solution contains 0.100 mg/ml of nickel.

(b) *Standard cobalt solution.* Dissolve 0.404 g of $CoCl_2 \cdot 6H_2O$ crystals in 100 ml of 0.1 M hydrochloric acid in a volumetric flask. This solution contains 1.00 mg/ml of cobalt.

(c) *Ammonium thiocyanate.* Dissolve 50 g of the salt in 100 ml of aqueous solution.

(d) *Dimethylglyoxime.* Dissolve 2 g of the solid in 200 ml of ethyl alcohol.

B. Prepare a Column Such as That Shown in Fig. 10.2.. (Note 1) Place a small plug of glass wool in the constricted end to support the resin. This end is also fitted with a small piece of rubber or plastic tubing and a screw clamp in order to regulate the flow of the solution through the column.

Place about 1 g of a strong-base anion resin (Note 2) in a 50-ml beaker and add about 10 ml of distilled water. Stir the resin to form a slurry and then add the slurry to the column using a "pipet" such as that shown in Fig. 10.3 (Note 3). Allow the resin to settle and the water to flow through the column and then add more slurry until the length of the column of resin is 4 to 5 cm (Note 4). Be sure that there are no air pockets in the column. Now add the small portions of 9 M hydrochloric acid to the resin and allow this to flow over the resin until 4 to 5 ml of acid has been used. (The resin will shrink somewhat and darken slightly on treatment with the concentrated acid.) Make two small marks, 1 cm apart, on the

[4] E. B. Sandell, *Colorimetric Determination of Traces of Metals*, 3rd ed., Interscience Publishers, Inc., New York, 1959.

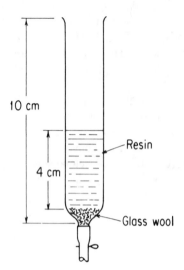

Fig. 10.2 Small ion-exchange column.

Fig. 10.3 Small "pipet" and syringe.

side of the column above the resin and adjust the screw clamp until the time required for the solution to flow this distance is 1 to 2 min.

C. Secure from the Instructor a Sample of a Solution Containing Nickel and Cobalt in 9 M Hydrochloric Acid. (Note 5) Using a 0.25 micropipet (250 λ) (Note 6), carefully add 0.25 ml of the solution to the resin column. Place the tip near the top of the resin bed and add the solution slowly so as not to stir up the bed. Place a clean 250-ml volumetric flask below the column to receive the solution. Allow the solution to seep into the resin and note the blue band (cobalt) at the top of the column. Then add slowly, about 1 ml at a time, 4 ml of 9 *M* hydrochloric acid, allowing each portion to settle into the resin before addition of the next portion. Note that the cobalt band spreads somewhat during the washing, but does not come off the column. Near the end of the last wash, catch a drop of the solution on a spot plate (or in a small beaker), add a drop of concentrated ammonia, and then add a drop of dimethylglyoxime. No red color should be formed, showing that the nickel has been completely removed.

Set aside the flask containing the nickel and replace it with a 50-ml volumetric flask to receive the cobalt. Add to the column about 4 to 5 ml of 1 *M* hydrochloric acid in successive 1-ml portions. Note that the cobalt band quickly begins to move down the column and observe the color of the drops as the cobalt is removed. (Why is the color first blue, then pink?)

D. Into Five 100-ml Volumetric Flasks Pipet 1-, 2-, 3-, 4-, and 5-ml Portions of the Standard Nickel Solution. Use a sixth flask for the blank. Add to each flask 1 ml of saturated bromine water, 4 ml of concentrated ammonia, 35 ml of 95% ethyl alcohol, and 20 ml of the dimethylglyoxime solution. Dilute to the 100-ml mark and measure the absorbance of each solution, using the blank as a reference. (Do not wait longer than $\frac{1}{2}$ h before making the measurements.) A wavelength of 450 nm is employed with a spectrophotometer, or a green filter with a filter photometer. To the volumetric flask containing the nickel unknown add 2.5 ml of bromine water, 14 ml of concentrated ammonia, 88 ml of 95% ethyl alcohol, and 50 ml of dimethylglyoxime solution. Dilute to the mark and measure the absorbance.

E. Into Five 100-ml Volumetric Flasks Pipet 1-, 2-, 3-, 4-, and 5-ml Portions of the Standard Cobalt Solution. Use a sixth flask for the blank. Add to each flask 10 ml of ammonium thiocyanate solution and 50 ml of acetone. Dilute each solution to the 100-ml mark and measure the absorbance at 625 nm if a spectrophotometer is employed, using the blank as a reference. To the volumetric flask containing the cobalt unknown add 1 *M* ammonia until the solution is only slightly acidic and then add 5 ml of ammonium thiocyanate and 25 ml of acetone. Dilute the solution to the mark with water and measure the absorbance.

F. Plot Absorbance vs. Concentration for Both the Nickel and Cobalt Standard Solutions. From the absorbances of each unknown solution calculate the concentrations in milligrams per milliliter of each metal.

A second portion of the unknown can be separated and determined if desired (consult the instructor).

Notes

1. This can be made by drawing down a piece of 9- or 10-mm-o.d. tubing. It may be furnished by the instructor. A 5-ml graduated pipet constricted near the 5-ml mark also makes a very convenient column. It can be cut off to any length desired.

2. Dowex 1, 8% cross-linked, 50 to 100 mesh, analytical grade, is suitable. Amberlite IRA 400 is a similar resin.

3. This can be made by drawing down a piece of 8- or 9-mm-o.d. tubing and slightly constricting the other end. A 5- or 10-ml syringe fitted with a rubber stopper is used to fill and empty the "pipet".

4. The capacity of such resins is usually of the order of 3 meq/dry gram. The column here is about 1 ml in volume, or approximately 0.4 g. Since the equivalent weights of both cobalt and nickel are about 29, this quantity of resin will hold about 35 mg of either of the metals. Only a few milligrams are actually added, so the column is operated at low "loading."

5. The solution can be about 0.05 M in nickel and about 0.1 M in cobalt.

6. Such pipets are normally calibrated "to contain," and should be rinsed once or twice with 9 M hydrochloric acid.

11

Solvent Extraction

The principles of solvent extraction are discussed in Chapter 16 of the text. Two experiments are presented here. The first is a simple extraction, illustrating the determination of the distribution ratio of a solute between two solvents. The second involves the separation of three metals using the Craig countercurrent distribution method.

EXPERIMENT 11.1. **Determination of the Distribution Ratio of Phthalic Acid between Water and *n*-Butanol**[1]

In this experiment the distribution ratio of phthalic acid between water and *n*-butanol will be determined. For the equilibrium

$$\text{Phthalic acid (water)} \rightleftharpoons \text{phthalic acid (}n\text{-butanol)}$$

this ratio is

$$D = \frac{[\text{phthalic acid}]_{n\text{-butanol}}}{[\text{phthalic acid}]_{\text{water}}}$$

[1] This experiment was designed by Hubert L. Youmans of Western Carolina University, who has kindly given us permission to use it here.

The brackets, as usual, represent molarities. However, the concentrations can be expressed as weight/volume since solute is the same in the numerator and denominator. The aqeous phase is 1 *M* in NaCl. This improves the phase separation and establishes an ionic strength of unity.

Procedure

Preparation of Solutions

(a) *1.0 M NaCl.* Prepare 300 ml of 1.0 *M* aqueous NaCl by dissolving 17.5 g of the salt in 300 ml of solution.

(b) *Solvents.* Place in a separatory funnel the 300 ml of 1.0 *M* NaCl and 250 ml of *n*-butanol. Shake the funnel thoroughly and separate the two phases.

(c) *Standard 0.1 M NaOH.* Prepare 1 liter of 0.1 *M* NaOH and standardize it against pure potassium acid phthalate (Experiments 3.1 and 3.3). The base may be standardized to only three significant figures.

Equilibration. Note: All extractions and titrations should be done in duplicate. Weigh 0.35 g of finely ground phthalic acid to three significant figures and place it in a 125-ml separatory funnel. Add 50 ml of the *n*-butanol (saturated with 1 *M* NaCl) to the funnel and dissolve the phthalic acid. Then add 60 ml of the 1 *M* aqueous NaCl (saturated with *n*-butanol) and equilibrate the system by thoroughly mixing for about 5 min. Allow the phases to separate; then drain them into separate 100-ml graduated cylinders. Record the volumes of the two phases.

Titration of the Aqueous Phase. Transfer the aqueous layer to a 250-ml Erlenmeyer flask. Add 3 drops of phenolphthalien indicator and titrate dropwise with standard base. Determine the blank correction by titrating 60 ml of the 1 *M* aqueous NaCl (saturated with *n*-butanol) that was not used in the extraction.

Titration of the Organic Phase. Transfer the organic layer to a 250-ml Erlenmeyer flask. Add 3 drops of phenolphthalein indicator and titrate slowly with standard base. This titration must be done slowly and with constant swirling because the phthalic acid is being extracted from the alcohol with the NaOH solution. When the aqueous layer permanently turns pink you have reached the end point. Determine the blank correction by titrating 50 ml of the *n*-butanol (saturated with 1 *M* aqueous NaCl) that was not used in the titration.

Calculations. Calculate the distribution ratio for the system. Report an average *D* to two significant figures. (Why are no more than two significant figures justified?) Calculate the mass balance for the experiment, that is, the weight of phthalic acid found in the two layers relative to the weight taken for the extraction.

EXPERIMENT 11.2.   Separation of Iron, Cobalt, and Nickel as the Thiocyanates[2]

In this experiment iron, cobalt, and nickel are separated by extraction of the thiocyanates with methyl isobutyl ketone. The distribution coefficients are sufficiently different so that the elements are separated in a 16-transfer distribution, the thiocyanates of the metals are brightly colored, and the concentrations can be determined by photometric measurements. The one disadvantage of the system is that the distribution coefficients of cobalt and iron vary with the ratio of thiocyanate to metal and thus change as the metal is distributed through the various tubes. Hence neither iron nor cobalt will distribute in the various tubes according to the simple binomial distribution worked out in Chapter 16 of the text. It is instructive to compare the experimental and theoretical distribution curves to see the effect of this complicating factor.

Procedures

Preparation of Solutions and Reagents

  (a)   *0.35 M ammonium thiocyanate.* Dissolve 26.6 g of ammonium thiocyanate in 1 liter of solution.

  (b)   *Methyl isobutyl ketone.* Obtained from Carbide and Carbon Chemical Co.—boiling point 117 to 119°C.

  (c)   *Standard iron solution.* Weigh accurately about 100 mg of pure iron wire (Note 1) in a 250-ml beaker. Add 7 ml of perchloric acid and a few drops of nitric acid. Heat to dissolve the metal and evaporate to strong fumes of perchloric acid. Cool, dilute somewhat, and transfer to a 100-ml volumetric flask. Dilute to the mark and mix.

  (d)   *Standard cobalt solution.* Weigh into a beaker approximately 150 mg of anhydrous iron- and nickel-free cobalt sulfate. Dissolve the salt in 100 ml of water and transfer to a 250-ml volumetric flask. Dilute to the mark and mix thoroughly.

Equilibration of Solvents. Remove any iron present in the 0.35 *M* ammonium thiocyanate solution (impurity in the ammonium thiocyanate used) by shaking the solution with 20 ml of methyl isobutyl ketone, allowing the mixture to separate, and removing and discarding the methyl isobutyl ketone layer. It may be necessary to repeat this operation. Equilibrate the solutions to be used in the following man-

---

[2]This experiment was worked out by Harvey Diehl, Iowa State University, Ames, Iowa, who has kindly given us permission to use it here. Diehl also recommends a simple experiment to illustrate the operation of the Craig apparatus. This is the distribution of methyl orange between isobutyl alcohol and water having a *p*H of 1.75. The distribution coefficient of methyl orange is constant (1.54) throughout the distribution, a good method of analysis is available, and the experimental and theoretical curves agree closely.

ner. Mix 1 liter of 0.35 *M* ammonium thiocyanate and 1 liter of methyl isobutyl ketone. Shake the mixture vigorously in a separatory funnel and allow to stand. During the next 10 min shake the mixture vigorously once or twice. Finally, allow the mixture to stand 1 h to come to room temperature and then separate the phases.

Preparation of Iron-Cobalt-Nickel Solution for Analysis. The solution containing the iron, cobalt, and nickel salts for the separation should contain an excess thiocyanate over that required for the metals present. A satisfactory mixture for analysis can be prepared in the following manner (Note 2). Transfer to a beaker 0.90 g of nickel nitrate, $Ni(NO_3)_2 \cdot 6H_2O$. Place in the same beaker 0.05 g of anhydrous cobalt sulfate and 1 mg of iron (1.0 ml of the standard iron solution prepared as described above). Add to the beaker 0.50 g of solid ammonium thiocyanate. Add 50 ml of 0.35 *M* ammonium thiocyanate solution previously with methyl isobutyl ketone.

Craig Distribution. Proceeding from right to left, number the tubes of the Craig apparatus successively 0, 1, 2, 3, . . . , 16. Into tubes 1 though 16 place 0.35 *M* ammonium thiocyanate, introducing the solution through tube E (Fig. 11.1). Fill each tube with sufficient solution so that some liquid will flow away through tube B when the apparatus is tilted with tube A vertical; that is, when the apparatus is tilted to the vertical position, the volume of the ammonium thiocyanate solution in each tube is determined by the pour-off point (junction of arm B with tube A). Tilt the apparatus through 90° (as far as it will go), causing the liquid above the pour-off point to flow into the auxiliary transfer chamber. Return the apparatus to the original horizontal position and allow the liquids a few moments to drain. Repeat this procedure twice. This will adjust the volume of 0.35 *M* ammonium thiocyanate solution in the first three tubes to exactly the volume of the end section of A; that is, each tube will be filled to the pour-off point. The remaining tubes will level off in the course of the subsequent operation.

In tube 0 pipet an appropriate volume of the iron-cobalt-nickel thiocyanate solution. Incline the apparatus to almost the vertical position and add equilibrated 0.35 *M* ammonium thiocyanate solution to fill tube 0 exactly to the pour-off point.

Add an appropriate volume of equilibrated methyl isobutyl ketone from a graduated cylinder to tube 0. Tilt the apparatus back and forth from about 30° above the horizontal to slightly below the horizontal; do this in such a manner that effective churning of the two liquid phases results, but not so vigorously that liquid is ejected through B or E. Rock the apparatus in this manner 20 times. Incline the apparatus to a position about 45° above horizontal and fix it in this position. Leave the apparatus in this position until the two phases have completely separated.

After the phases have separated cleanly, carefully tilt the apparatus as far back as the stop will permit. This will cause the methyl isobutyl ketone to flow into the auxiliary transfer chamber C; during this operation no liquid should be allowed to flow through arm D. Slowly and carefully return the apparatus to the horizontal position. During this operation liquid may rise from tube 0 into arm B,

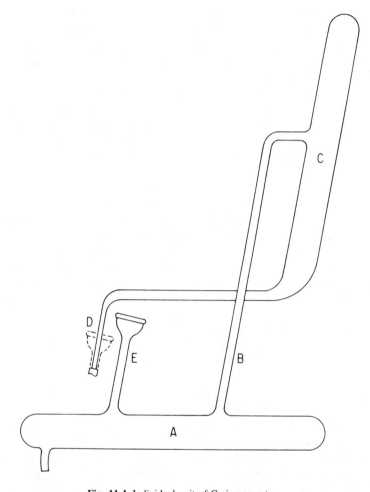

Fig. 11.1 Individual unit of Craig apparatus.

but the column will usually break and flow back into tube 0; this rise of liquid in arm B can be prevented by quickly tilting the apparatus slightly toward the vertical position.

After the transfer of methyl isobutyl ketone from tube 0 to tube 1 has been completed, allow the liquids a few moments to drain. Again add the same volume of ketone to tube 0. Rock the apparatus as before to bring the immiscible liquids into thorough contact (in tubes 0 and 1 simultaneously). Then separate the phases as before, transfer as before, again charge tube 0 with ketone, and so no. Make 16 such cycles altogether. This will mean a total of 136 individual extractions, that is, one on the first cycle, two on the second, three on the third, and so on.

Clean seventeen 100-ml volumetric flasks (Note 3) and number them from 0 to 16. Into each of the flasks 5 through 12 (those which will contain the cobalt), add 10 g of solid ammonium thiocyanate. Drain the total liquid from tubes 0 through 16 into the corresponding flasks. To flask 0 only, add the same volume of methyl isobutyl ketone as used throughout. Replace the clamps on tubes 0 through 16 and fill each tube with methanol to the pour-off point. Rock the apparatus two or three times and drain the methanol into the corresponding flasks.

Fill each flask to the mark with methanol. Some contraction occurs on mixing and it will be necessary to add more methanol after mixing. Finally, fill the flasks to the mark with methanol and mix well by shaking. Completely homogeneous solutions should result. Analyze each solution as described below. Clean the Craig apparatus immediately by rinsing it twice with distilled water and once with methanol.

Spectrophotometric Measurements of Iron and Cobalt. Following the Craig distribution of the iron-cobalt-nickel mixture, the nickel should be in tubes 0 and 1; the cobalt should be in tubes 5 through 12; and the iron in tubes 12 through 16. The determinations of iron and cobalt can be made conveniently by photometric measurements; indeed, the quantities specified in the experiment above were chosen to yield solutions containing suitable amounts of the metals for photometric measurement. Nickel cannot be determined with any accuracy as the thiocyanate, but fortunately little interest resides in it, for it remains in tube 0 during the distribution, with possibly a little mechanical carryover in tube 1.

The photometric determinations may be made with a spectrophotometer using a wavelength of 493 nm for iron and 625 nm for cobalt.

Spectrophotometric Determination of Iron. Prepare a secondary iron standard using the standard iron solution prepared as described in (c) on page 173. Carefully pipet 5.0 ml of the primary standard iron solution into a 250-ml volumetric flask, dilute carefully, and mix well. Calculate the iron content of this secondary standard solution, which should contain about 0.025 mg of iron/ml.

Prepare four or five solutions of iron from this secondary standard by pipeting various volumes into 50-ml volumetric flasks in such a manner that the final iron concentration will vary between 0.04 and 0.6 mg of iron/100 ml. To each flask add 15 ml of 0.35 M ammonium thiocyanate and 15 ml of methyl isobutyl ketone. Add 1 drop of concentrated sulfuric acid. Dilute to the mark with methyl alcohol and mix thoroughly as the dilution is made. Measure the absorbance at 493 nm. Prepare a calibration curve by plotting absorbance against concentration.

Determine the iron in tubes 13 through 16 following the same procedure.

Spectrophotometric Determination of Cobalt. Using the standard cobalt sulfate solution prepared as described in (d) on page 173, prepare a series of secondary cobalt standards. Pipet 5.0-, 10.0-, 15.0-, 20.0-, and 25.0-ml portions of this solution into 50-ml volumetric flasks. Dilute to the mark and mix well. This will give five secondary standard solutions. Pipet 10.0-ml portions of these solutions into 50-ml volumetric flasks. Add 10 g of ammonium thiocyanate and 15 ml of methyl isobutyl ketone. Dilute with methyl alcohol to the mark and mix well. Some contraction occurs on mixing; add more methyl alcohol to bring the meniscus to the mark and mix well. The final solution should contain 0.6 to 6 mg of cobalt/ 100 ml.

Prepare a calibration curve by plotting absorbance against concentration. The slope of this curve gives (if the concentration is expressed on a molar basis) ϵb; this will be about 1730 if $b = 1$ cm.

Determine the cobalt in tubes 5 through 12 in the same manner.

Iron absorbs somewhat at 625 nm, but even in tube 12 the amount of iron and its absorbance is so small as to introduce an error less than 1% in the cobalt determination.

Calculation and Expression of Results. Calculate the weight of cobalt and iron in each tube using the absorbances observed at the respective wavelengths and the calibration curves prepared. For each of the metals calculate the total weight present in all the tubes (5 through 12 for cobalt, 13 through 16 for iron); also, for each metal calculate the fraction found in each tube. On a graph plot the fraction in each tube against the tube number (as abscissa). The curve for cobalt is asymmetric, the cobalt moving out more rapidly than it should.

It is instructive to plot on the same graph with the experimentally found distribution curve a theoretical distribution curve for cobalt for some average value of the distribution coefficient as simply defined. A value of $K_D = 1.4$ gives a curve with a peak in tube 9. Look up the binomial coefficients for $n = 16$. A comparison of the experimental and theoretical curves shows the effect of the different character of the distribution law applying to the cobalt-thiocyanate system.

Notes

1. The iron solution can be prepared by weighing about 0.86 g of reagent grade iron(III) ammonium sulfate (uneffloresced) and dissolving it in 1 M perchloric acid in a 100-ml volumetric flask. Alternatively, iron(II) ammonium sulfate can be weighed and oxidized with hydrogen peroxide. Consult the instructor.

2. The instructor may wish to give you an unknown solution here and have you report the total quantities of iron and cobalt in the solution.

3. This size of flask is chosen for a starting volume of about 30 ml of the metal solution. It should be chosen in terms of the appropriate volume for the particular apparatus at hand.

Index

Page numbers followed by
  (t) refer to tables
  (n) refer to notes
  (f) refer to figures

178